The Netherlands and the Oil Crisis

The Netherlands and the Oil Crisis

Business as Usual

Duco Hellema
Cees Wiebes
Toby Witte

AMSTERDAM UNIVERSITY PRESS

The translation of this publication was funded by the Netherlands Organisation for Scientific Research (NWO).

The Netherlands and the Oil Crisis: Business as Usual is a translation of *Doelwit Rotterdam: Nederland en de oliecrisis*, Den Haag: Sdu, 1998.
English translation: Murray Pearson

Cover illustration: © Freek van Arkel/Hollandse Hoogte

Cover design: Sabine Mannel, N.A.P., Amsterdam
Lay-out: Adriaan de Jonge, Amsterdam

ISBN 90 5356 485 3
NUR 697

Contents

Foreword

October 2003 marked the 30th anniversary of the Arab oil embargo levied against The Netherlands as a 'punishment' for its pro-Israeli stance in the October War. On October 6, 1973, Egyptian and Syrian troops attacked Israel in an attempt to regain the land occupied by Israel since 1967, and for several days the Israeli army had its back against the wall. In The Netherlands, the first reports of the war aroused great concern: Israel must be helped as in 1956 and in 1967.

The Dutch government led by Prime Minister Joop den Uyl had been in power in The Netherlands since May 1973, a coalition consisting of, on the one hand, three progressive parties, the Dutch Labour Party (*Partij van de Arbeid*, PvdA) and two smaller parties: the progressive-liberal D'66 and the Radical Party (PPR), and on the other hand, the Catholic People's Party (the *Katholieke Volkspartij*, KVP) and the protestant Anti-Revolutionary Party (ARP). After the outbreak of the October War, the Den Uyl Cabinet left no doubt as to its pro-Israeli sympathies, making it clear in a governmental statement that it held Egypt and Syria responsible for initiating hostilities and for unilaterally violating the truce. In the European Community, too, The Netherlands took a more emphatically pro-Israeli stand than did other member states, so much so that for a while The Hague found itself isolated.

Nor was this merely a matter of words. Under conditions of strict secrecy, a considerable quantity of ammunitions and spare parts was sent to Israel, an extensive military operation in Dutch terms, which it has long been maintained was undertaken without the knowledge of the Cabinet. This political and military support for Israel would subsequently be given as the reason for an oil embargo levied against The Netherlands. Yamani, the Saudi Minister responsible for oil matters, himself declared that this was the main motive for the embargo.

Yet the affair of the oil embargo was by no means merely a response to the help lent to Israel. The oil crisis was also part of, or rather an expression of, an intense power struggle in the international oil sector. The radical Arab oil producers were intent on breaking down the traditional relations in this sector in which The Netherlands occupied an important position. It was the homeport of Shell, one of the largest of the oil multinationals. Furthermore, Rotterdam was a crucial switch-point in the whole circuit of the processing and distribution of oil in Western Europe. An embargo against The Netherlands seemed to affect half of North-West Europe.

In various respects, the oil crisis was a first test case for the Den Uyl Cabinet, for it presented enormous problems, not only of foreign policy but also with regard to domestic and socio-economic affairs. In the arena of international politics, the oil crisis demanded that fundamental choices be made concerning relations between North and South, the American-European relationship and relations within the European Community.

The oil crisis had a huge influence on Dutch domestic politics. The Central Planning Bureau predicted a marked rise in unemployment, slackening economic growth, increased inflation and possibly great damage to the port of Rotterdam and Dutch business life. For on paper, as one newspaper wrote a few days after the announcement of the embargo, turning off the oil tap was nothing short of a national disaster. The Dutch public was confronted with the prospect of Sundays without cars, of petrol rationing and restrictions on the use of electricity.

What above all prompted us to write this book was the fascinating and at the same time complex totality of the oil crisis. In addition to which, this crisis suddenly placed The Netherlands centre stage in the theatre of international politics. The oil embargo focused all eyes on The Hague. So far, relatively little has been written on the role of The Netherlands during the oil crisis. Several studies have appeared, but an extensive study covering the whole range of different aspects has been lacking. More curious is the fact that no one has hitherto undertaken a thorough investigation of the archives of those ministries most involved in the oil crisis.

Thanks to the Dutch Freedom of Information Act, we were allowed ample access to the most restricted records that had a bearing on the crisis; which is to say that those ministries closely involved in the crisis – Foreign Affairs and Economic Affairs – made their records freely available. Abroad too, specifically in the United States, we were granted access to relevant archival documents, often for the first time.

Naturally, a number of questions remain unanswered. On certain points the archives contain no information, such as, in the matter of secret weapons deliveries to Israel. Nor were interviews always capable of filling these lacunae. Furthermore, the oil companies involved declined to allow us access to their company archives, because of which, in part, we have had to set ourselves several limits and have deliberately left (indeed, have had to leave) some aspects unconsidered.

In addition, we pay scant attention to the financial-monetary aspects of the Dutch position or to the long-term consequences for the Dutch economy. Nevertheless, this study does, in our view, embrace several new points of view on Dutch foreign policy and, not least, on the policy of the Den Uyl Cabinet.

In the end, we decided to write a case study focusing mainly on the political actions of the government, concentrating mainly on those ministers and ministries most significantly involved. The construction of the book is such that we try in each chapter to deal with a particular aspect of the oil crisis: Dutch Middle East politics, Dutch European politics, domestic measures, and so on.

We are most grateful to the following (archive) assistants and civil servants who provided help: Francien van Anrooy and Sierk Plantinga of The National Archives; Fred van den Kieboom and Radjen Gangapersadsing at the Cabinet Office; Hans den Hollander, Henja Korsten, Peter van Velzen, Marco Verhaar and Ton van Zeeland at the Foreign Ministry; Sam Martijn of the Central Archives Depot at the Ministry of Defence; Th.J.N. Knops, Henrietta Kruse and J. Zuurmond at the Ministry for Economic Affairs; Ella Molenaar, Monique van der Pal, Cees Smit, Willeke Tijssen and Mieke IJzermans of the International Institute for Social History and Jaap van Doorn and Maarten van Rijn at the Ministry of Justice.

We would also like to thank those individuals involved at different stages who have been prepared to read (parts of) the manuscript and offer constructive criticism. These were: F.E. Kruimink (then Co-ordinator of the Dutch Intelligence and Security Services), J.P. Pronk (Minister of Development Cooperation), A. Stemerdink (Under-Secretary of Defence), M. van der Stoel (Minister of Foreign Affairs), H. Vredeling (Minister of Defence), W.Q.J. Willemsen (Secretary of the Co-ordination Group for Oil Crisis Management and of the Ministerial Commission on the Oil Crisis) and G.A. Wagner (Chief executive of the Royal/Shell Group). We also wish to thank Paul Aarts (University of Amsterdam) for his advice; and to extend thanks to all those who were ready to give us their time in interviews or to provide written answers to our questions. Of course, we as authors take full responsibility for the final text of this book.

Amsterdam, September 2004

Duco Hellema, Cees Wiebes and Toby Witte

1

War in the Middle East

On 6 Ocober 1973, large numbers of Egyptian and Syrian military units crossed the frontiers with Israel that had held since 1970. Around 240 Egyptian warplanes crossed radar installations. At the same time, some 1800 artillery and mortar positions opened up along the whole front and 700 Syrian tanks attacked the Golan Heights where the Israeli land forces had only been able to deploy some 150 tanks. Although reports had already been circulating throughout the summer of an Egyptian-Syrian attack, the Israeli army command appeared to be caught by surprise. It seems that they were only convinced that the threat was serious a few days before the actual outbreak of the war. The possibility of a pre-emptive strike was briefly considered, but there was insufficient time for the necessary preparations. And furthermore, Israel would then be branded in international opinion as the aggressor. The decision therefore, as the Dutch Ambassador G.J. Jongejans reported to The Hague, was to wait whilst at the same time 'seeking the full moral and political advantage' of that restraint.[1]

The question is whether the aggressors really had set themselves the aim of 'wiping Israel from the map'. Possibly their intention was merely to realise limited military objectives and to cause an international crisis which would make the Great Powers realise that continued political impasse was unsustainable. Whatever the case, the Arab advance was an impressive success, and the Israeli military situation rapidly became serious. Within a very short time, the Egyptian forces had crossed the Suez Canal and broken through the Israeli line of defence. The Egyptian Army was able to re-take most of the Suez East Bank, occupied by Israel ever since the 1967 war, while at the same time the Syrian army succeeded in occupying a large part of the strategically important Golan Heights. It seemed that a real disaster for Israel was taking shape.

After about four days, however, events began to turn. Israel managed to halt the offensive and began its own counter-attack. Israeli forces managed to regroup on the Golan, and on October 10, tank units broke through the Syrian defences, bringing Damascus within range of Israel's artillery. Tel Aviv decided, however, not to pursue this course, since the Soviet Union had made it clear that any attack on Damascus would not be tolerated. In addition, further advance would be likely to incur unacceptable losses and would also run the risk that Jordan might become more actively involved. In the Sinai, the Egyptian army was managing for the time being to stand its ground, but during the night of October 15, Israeli tank units crossed the Suez Canal with the aim of isolating the Egyptian 3rd army. The plan worked, and on October 21 this army corps was almost completely cut off from the outside world.[2]

On October 16, when it became apparent that Egypt and Syria were in deep trouble, Soviet premier Alexei N. Kosygin flew to Cairo to urge the Egyptian president Anwar Sadat to call a cease-fire. He showed Sadat satellite photos of the Israeli advance, and on October 18, when the Israeli army had established a bridgehead on the western bank of the Suez Canal, Sadat agreed. Leonid Brezhnev, the Soviet leader, informed Kissinger and Nixon (who were at the time totally preoccupied by the Watergate affair) of Sadat's willingness, and because Washington did not immediately respond, Brezhnev invited Kissinger to Moscow for further talks. This delay allowed Israel time to advance further against Egypt.[3]

On October 22, the Security Council adopted a resolution calling for a cease-fire. Although the two sides accepted this resolution, the fighting in fact continued. In the night of October 24–25, the Security Council again called for a cease-fire and further demanded that the belligerent parties withdraw to the positions held on 22 October. International tension increased. The Soviet Union threatened direct military intervention if the Israeli advance were not halted. Washington reacted on October 25 by putting into operation Defense Condition 3 (*DEFCON III*) which meant that the American armed forces were put on a higher alert, including the announcement of a *nuclear alert*.[4] To the annoyance of West European countries, this also involved the American troops in Europe, even though there had been no discussion of this within NATO. Emotions were quickly calmed, however, when Washington withdrew *DEFCON III* on October 26. One day later, the first meeting took place between Israeli and Egyptian officers. Three days later an agreement was reached on the exchange of prisoners of war.[5]

The October war was not the first confrontation between Israel and its neighbouring Arab states. There had been wars in 1948, 1956 and again in 1967, all of them decided in Israel's favour. In the Six-Day War, in particular, Israel had succeeded in considerably expanding its territory, taking in the Sinai desert (which led to the closure of the Suez Canal), the Gaza strip, the Golan Heights, and the West Bank of the Jordan. The end of the war initiated a period of protracted and fruitless diplomatic manoeuvrings in search of a peace accord, the starting point for which would necessarily have to be an end to this Israeli territorial expansion. To this end, on 22 November 1967, the Security Council adopted resolution 242, a resolution which in subsequent years was to give rise regularly to diplomatic differences of interpretation, even within the EC. While the English version called for 'withdrawal of Israel's armed forces from territories occupied in the recent conflict' the French version demanded that Israel withdraw from *the* territories occupied ('retrait des forces armées israeliennes des terrritoires occupés').

The failure to find a political solution to the Middle East conflict ensured continuous tension in the area from 1967 to 1973, with the differences between the Soviet Union and the United States playing an increasingly important role. Time and again, hostilities flared between Israel and an Egypt enjoying large-scale military and economic support from Moscow. In the summer of 1970, after long and delicate negotiations, a cease-fire was agreed, but it proved impossible to reach agreement over a peace accord. In the Arab world, this impasse served to increase frustration. In January 1973, Sadat warned that a new war was beginning to look inevitable unless a political solution could be found soon. On October 6, he was vindicated.

The outbreak of the October War brought the two Great Powers, the Soviet Union and the United States, unexpectedly and sharply into conflict. From the early 1970s on, relations had improved between the Soviet Union and the USA. In May 1972, Brezhnev and Nixon had met in Moscow and jointly signed the SALT-I treaty. In the meantime, the Conference over Security and Cooperation in Europe (CSCE) had been set up. It seemed that a new period of détente had begun.

Egyptian and Syrian dissatisfaction with the situation in the Middle East was well-known. Rumours had circulated earlier of a possible Egyptian-Syrian attack on Israel, and in January the joint armed forces of Egypt, Syria and Jordan were put under the command of the Egyptian Minister for War. Nonetheless, most informed opinion held that the three nations possessed insufficient military resources to see such a venture through successfully.[6]

Early in 1973, Moscow decided to supply Egypt with extra military material whilst at the same time, in Cairo, continuing to press for a diplomatic solution. Moscow's strategy failed, for on October 4 the Soviet leaders were informed of Egypt's intention to attack Israel within a few days. Immediately, the Kremlin sent transport planes to both Cairo and Damascus to collect the families of advisors and diplomats. Washington, however, remained convinced that Sadat would not start a war.[7] Once war had broken out, both Washington and Moscow immediately set up an airlift. The Soviet airlift came into operation the third day of the war, in spite of the fact that the Soviet military leadership expected Arab successes to be of only a temporary nature. On October 9, the airlift was concentrated solely on Syria, since the Syrian military situation was rapidly deteriorating. Supplies were still modest: no tanks or aircraft, only fuel and ammunition; but on October 10, *Antonov-12* and the gigantic *Antonov-22* transport planes began flights to Egypt.[8]

On October 14, several days after the Soviets began provisioning and after a week of hesitancy and differences of opinion, the USA announced that it was beginning delivery of weapons to Israel. The American airlift ran via the Azores. In all probability, however, the Americans began supplying Israel earlier, albeit on a limited scale. For example, as soon as the war broke out, American supplies were redirected from West Germany.

The major West European countries were reticent in their response to the Middle Eastern war. Both French and British governments called for an arms embargo and refused to provide the belligerent parties with materials, an attitude which rapidly assumed an anti-Israeli complexion, at least partly as a result of statements from both countries over who was to blame. In addition to which, France continued to supply other Arabic countries, for example Libya, without specifying that such supplies should not find their way to Egypt or Syria.[9] West Germany adopted a much less outspoken approach. Bonn declared that it was not wholly in sympathy with Israel, but turned a blind eye to the movement of American supplies to Israel from West German soil. Within the European Community, not only did a majority seem disinclined to come out openly or actively in support of Israel, there was also unease over the role played by the Americans in the war and over the lack of consultation, particularly when Nixon put American troops on a high alert. In this regard, however, The Netherlands was the exception.

The Netherlands and the Middle East

As during the 1967 war, the first reports of the war in the Middle East caused great concern in The Netherlands. In its assessment of developments in the Middle East, The Netherlands had firmly allied itself with Israel over two decades. Originally, the Dutch had vacillated for some time before recognising the Jewish state, careful lest Islamic Arab countries should be antagonised in view of the problems with Indonesia.[10] But during the 1950s and 1960s, a 'special alliance' developed between the two countries. During the wars of 1956 and 1967, the Netherlands supplied Israel with military materials, though it should be added that policy in 1956 was heavily influenced by the desire to see Nasser's Egypt brought to heel.[11] The government also tried to offer Israel diplomatic support on various other occasions, in particular in the 1960s over the question of the Israeli-occupied territories.

In the diplomatic battles over the question of whether Israel should return *all* of the occupied territories to the surrounding Arab nations, The Netherlands always interpreted resolution 242 in such a way that the possibility of strategic 'border corrections' would remain open. It was frequently emphasised in The Hague that Israel had the right to *secure borders*. On the Palestinian question, too, The Hague came out in support of Israel: the position and status of the Palestinians were a humanitarian and not a political issue. In the General Assembly of the UN, The Netherlands voted against resolutions calling for the Palestinian people's right to self-determination.

The period in office of the Foreign Minister, W.K.N. Schmelzer (1971-1973), saw a cautious change of political direction. In the General Assembly in December 1972, The Netherlands backed the famous resolution 2949 which recognised the rights of the Palestinians as an inseparable part of the peace process, in spite of both Israeli and American dissent. In the same year, Schmelzer declared that border corrections were only possible if all parties accepted them. Inevitably, such views introduced an element of estrangement into Dutch-Israeli relations.[12] These developments were accompanied by a closer rapprochement with the Arab countries, a process already begun in the late 1960s.

During this period, although it became increasingly more difficult for the Foreign Ministry to consent to arms deliveries to Israel,[13] Dutch-Israeli military contacts were never completely broken. Israeli soldiers, for example, trained in The Netherlands in 1971 and 1972. Military instructors familiarised their Israeli colleagues with the lightly armoured personnel carriers, the M-113's; Israelis learned to shoot, drive, manoeuvre and navigate the M-113's, and for this purpose they received secret training at the Royal Engineers Training School for Navigation and Diving. An exercise involving crossing the Maas with an M-113 was held near Zaltbommel.[14] The M-113's were to be used in crossing the Suez Canal in October 1973. According to the Military Intelligence Service (MID) documents, the Israelis conducted their exercises in The Netherlands because 'in the circumstances instruction from the American side would have been too sensitive'.[15]

Shortly before the October War, The Netherlands and Israel were still cooperating in the modernisation of the cannons of Israel's Centurion tanks. A number of Israeli military personnel were given training at the Army tank workshops in Amersfoort. Parts for Centurions were also flown to Israel from the Soesterberg military airfield,[16] a rather remarkable transaction since the new (PvdA) Minister, M. van der Stoel, had decided on August 30 that all military deliveries to states immediately involved in the Israeli-Arab conflict must cease.[17]

Support for Israel

On October 6, at the request of the Israeli Foreign Ministry, the Dutch Ambassador in Jerusalem sent a communication to The Hague that there were 'massive concentrations of Egyptian and Syrian armed forces in attack positions' gathered on Israel's borders. From New York it was also reported that UN observers were seeing 'strong indications' that Syria and Egypt wanted to embark on acts of war. It was assumed in New York that the intention was probably to achieve limited military objectives in order to be able subsequently to exploit the political situation. The Ambassador in Jerusalem was of the same view, maintaining this assumption even after the outbreak of the war. It was further assumed that Israel would rapidly push back the invaders through its supremacy in the air, an assessment that soon proved overly optimistic.[18]

In spite of the political shifts of the preceding years, The Netherlands came out 100% behind the Israeli cause. At first, The Hague – including the Dutch Foreign Ministry – was uncertain of the situation in the Middle East. On Sunday, October 7, Van der Stoel in fact was not prepared to comment. In the meantime, it was clear that Egypt wanted to petition the General Assembly and did not want the Security Council to intervene. Van der Stoel's preference, on the contrary, was for the Security Council to be brought in. The Permanent Representative in New York, R. Fack, was nevertheless instructed not to oppose a debate on the Middle East in the General Assembly.[19]

On October 8, a high-level discussion of the conflict took place at the Foreign Ministry between J.M. den Uyl, Justice Minister A.A.M. van Agt, Minister of Economic Affairs R.F.M. Lubbers, Van der Stoel and Defence Minister H. Vredeling. At that moment, Israel was in serious trouble. As reports of this meeting reveal, it was decided to call for a cease-fire as soon as possible on conditions acceptable to both sides, preferably on the basis of a *restoration of the status quo ante.*[20] The co-operation of the EC countries must be enlisted to prevent any Arab resolution which labelled Israel as the 'aggressor' being passed in the General Assembly. In any case, The Netherlands would vote against any such resolution, and would press for a quick meeting of the Security Council. In brief, it was decided during this consultation to lend all possible diplomatic support to Israel.[21]

In accordance with the conclusions of this consultation, the Permanent Representative at the UN, R. Fack, was thoroughly briefed the same day. Above all, he was to oppose any resolution which condemned Israel as the aggressor. He was also instructed to vote against any resolution that demanded the implementation of resolution 242, since the situation was now very different following the breaching of existing borders. The Security Council was the appropriate organ for ending the conflict, to which end the first priority was suspension of the armed struggle, preferably on the basis of a restoration of the *status quo ante.*[22]

On Monday, October 8, the Dutch Foreign Ministry issued its first press statement on the war. According to this statement, it could be deduced from the reports of UN observers that Egypt and Syria had initiated the open violence. The government hoped that the Security Council could find a formula acceptable to both parties that would lead to a cease-fire.[23] On the afternoon of October 8, the Permanent Committee for Foreign Affairs met in emergency session. At the end of this consultation, it was given out to the press that all political parties, with the exception of the communist CPN and the pacifist PSP, could endorse the government's position that restoration of the existing pre-war borders was essential.[24]

Meanwhile, under the instigation of the Second Chamber, a government statement was being prepared. This statement was worked out at the Foreign Ministry by the Department for International Organizations (DIO), a branch of the Directorate-General for International Cooperation (DGIS). This arose from the fact that, in the Netherlands at least, Middle Eastern policy had until that point always been a UN affair. The Director-General for Political Affairs (DGPA), D.W. van Lynden, disagreed with what he considered an excessively pro-Israeli statement.[25] This criticism of the policy pursued by the government, expressed by the DGPA, was to remain a significant factor throughout the entire period of crisis. Van Lynden, like a number of diplomats involved, continued to urge that this standpoint be modified to go some way to meet the Arab countries.

On October 9, the government statement was made public. The Cabinet, it said, had noted with consternation the resumption of the acts of war initiated by Syria and Egypt, as was evident from the reports of UN observers, among other sources. Egypt and Syria had thus unilaterally broken the truce that had held since 1970. The two assailants should therefore withdraw behind the armistice lines observed prior to October 6. The government called on the Security Council to try to achieve a political solution based on Security Council Resolution 242.

Through its choice of words, the Den Uyl Cabinet made it clear that it still stood firmly behind the interpretation of resolution 242 that the Arab countries considered pro-Israeli: Israel must withdraw *from occupied areas* (without the definite article). At the same time it was proposed that a political solution to the conflict had to be inseparably linked with a *just and fair solution to the refugee question*, meaning the question of the Palestinians.[26] What the declaration meant – as had been agreed in the ministerial discussions mentioned earlier – was support for Israel. In the event, the Dutch government statement was fairly generally supported in the Second Chamber, notwithstanding observations on the Palestinian question made by the PvdA and PPR. The PvdA leader, E. van Thijn, emphasised the need to strive for a solution to the Middle East conflict that would do justice to the political aspirations of the Palestinians.[27]

In New York, this government statement caused consternation among the Dutch Permanent Representation at the United Nations. Once Fack had scrutinised the statement he decided in consultation with his second man, subsequently Minister C. van der Klaauw, to give it as little publicity as possible. According to Fack, those in The Hague had been 'unwise' since The Netherlands could not, after all, 'maintain that a country trying to regain its own territory from a foreign occupier was committing aggression'. The position of conscientious balance and probity in the question of the Middle East, a position carefully constructed over past years, now threatened to collapse, Fack believed, like a house of cards.[28]

This was a first indication that his diplomats did not always subscribe to Van der Stoel's viewpoint.

But the government's position could obviously be bent more towards the Arabs. On October 10, Van der Stoel had an interview with the Ambassadors of Saudi Arabia and Tunisia and with the Egyptian temporary chargé d'affairs concerning the Dutch position. It appears that at this meeting the Foreign Minister emphasised the fact that the government had called for a cease-fire based on a situation acceptable to both parties. It was not for The Netherlands, argued Van der Stoel, to say what the conditions should be before a cease-fire could be reached, although he stated his preference for a restoration of the truce boundaries of August 1970. The three Arab diplomats were evidently satisfied with this clarification.[29] At that moment, it seemed that the two principles, 'restoration of the *status quo ante*' and 'a cease-fire acceptable to both parties', were evenly balanced. Two days later, however, during a European Political Cooperation consultation, the Dutch emphasis had shifted pre-eminently to the side of restoring the *status quo ante*.

This did not prevent The Netherlands finding itself rapidly isolated within the EC. It was announced in the government statement that The Hague would make its attitude better understood within the consultative process with the nine member states. It was soon evident, however, that this was no easy matter, since most EC countries were not inclined to offer Israel their support. France and Italy, in particular, and to a lesser extent the UK, seemed rather to choose the Arab side. In the case of France, this became apparent during a sitting of the Security Council, when the French delegate pointed out that the current fighting was taking place in areas that had been occupied by Israel since 1967. France sought a peaceful solution to the conflict on the basis of resolution 242 (i.e. no restoration of the *status quo ante*) and expressed the desire that the entire Middle East problem should be involved in such a settlement.[30] It was evident in The Hague that French sympathies inclined to the side of the Arab countries. The Security Council appeared paralysed for the time being because the USA would not desert the Israelis, and furthermore neither Israel nor the Arab countries put much stock in any pronouncement from the Council. The conflict was to be decided on the battlefield.

Military Support

As we said earlier, the first reports arriving in The Hague, both from Jerusalem and New York, were still fairly optimistic over Israel's military position. It was assumed that it was a limited military conflict in which Israel's military superiority would ensure a swift Israeli victory. On Monday the 8th, Ambassador G.J. Jongejans reported from Jerusalem that the Israeli Cabinet had the previous day authorised crossing the existing truce boundaries. 'As far as Israel was concerned, the war was actually already won.'[31] But it turned out to be far from as easy as that.

On October 7, the second day of the war, in a dramatic conversation the Israeli Ambassador to The Netherlands, C. Bar On, asked Minister Van der Stoel for military and material assistance. Undoubtedly, the Ambassador was fully aware that The Netherlands had assisted Israel with military material in the wars of 1956 and 1967. Bar On let it be known that Israel stood on the edge of the abyss and was desperately in need of ammunition and spare parts. The British and French governments had issued a ban on the export of arms to the warring parties as soon as the war broke out. American material support, to all appearances, was also very slow to get going during the first days of the war. Unlike Great Britain and France, the Dutch government did not ban the export of arms to the belligerents. According to the Ambassador, The Netherlands appeared to be the only remaining candidate for supplying Israel with the much-needed ammunition and spare parts.[32]

As Bar On recalls, Van der Stoel reacted rather coolly. Perhaps his caution was dictated by the initially optimistic reports from Jerusalem, but in any case he wanted to wait and see exactly how serious the Israeli situation was. Bar On remained in contact with Van der Stoel and with Premier Den Uyl and Vredeling, the Minister of Defence, throughout the following days.[33] The contacts with Vredeling were arranged by the PvdA Member of Parliament H. van den Bergh. Vredeling meanwhile had already intimated to him that he wished to send arms to Israel.[34]

As Minister of Defence, Vredeling played a central role in supplying arms to Israel. He has always stressed that his position was based on emotional, personal considerations. The events of the Second World War, the ex-resistance fighter later explained, must never be allowed to happen again. But despite all Vredeling's noble-minded aims, the fact was that The Netherlands was being discretely pressured by the Americans.

The background to this was that Foreign Minister Secretary Kissinger and Defence Minister James R. Schlesinger had clashed over extra arms deliveries to Israel. The initial American reaction was one of reservation, but apparently this caused considerable unrest in some quarters in Washington, given Israel's initially threatened military situation. On Sunday morning, the CIA operator at the American embassy in The Hague in charge of communications with headquarters in Langley, Virginia, received a *critic* from CIA headquarters. Such a coded telegram requires a response within a few hours. He therefore contacted the CIA Chief of Station in The Netherlands, Carlton B. Swift Jr., who had arrived in The Hague in the summer of 1973.[35]

Swift was instructed to approach the Dutch Cabinet to supply Israel with as many weapons and spares as possible. The *critic* emphasised that the political heads of the American Embassy had not been informed of these instructions. The *critic* that Swift received contained this brief request to the Cabinet to satisfy the Israeli requirements to whatever extent possible.[36] Swift carried out his brief in discrete fashion. On Sunday morning he contacted the Head of the Dutch Internal Security Service (the BVD), D. Kuipers, and the Intelligence and Security Co-ordinator of the Ministry of General Affairs, (the Prime Minister's Office) F.E. Kruimink, who later confirmed that he and Kuipers were unexpectedly called at home by Swift on Sunday, October 7, to discuss a matter of great urgency.[37] Swift's request found a receptive audience. Kruimink was to play an active role in the deliveries of arms.[38]

As we said earlier, a meeting on Monday, October 8, of the five Cabinet members most involved led to the conclusion that The Netherlands should support Israel. Both Van der Stoel and Vredeling deny that military support was discussed at this meeting.[39] Nonetheless, a remarkable incident occurred that same day. As the newspaper *De Telegraaf* reported, 'two days after the outbreak of the war', two Israeli transport planes arrived at Gilze Rijen airport. A note in the Den Uyl archive, written by Den Uyl himself, reads: 'Two days after the outbreak of the war in the Middle East, the Cabinet allowed several Israeli transport planes that had come to The Netherlands to fetch armaments to return empty-handed'.[40]

The journalist F. Peeters, who has written a book on the Dutch-Israeli military alliance, believes that the two aircraft actually left loaded.[41] Under-secretary for Defence Stemerdink confirmed that the two aircraft had indeed been loaded with American communication and detection equipment sent from West Germany. In all probability there were no Dutch materials sent; there had been at that stage inadequate preparation on the part of the Dutch.[42] In Vredeling's view, there had been no political permission for this procedure, and in any case he himself was not fully informed.[43] Stemerdink, however, was of the opinion that no Dutch permission was necessary, since it was a matter of American transport of materials, as had often occurred before. The Dutch government had nothing to do with it.[44] Stemerdink takes a rather laconic line here, since the aircraft that collected the American materials were from Israel, a country at war. In this connection, it is even more remarkable that Den Uyl noted that the *government* had allowed these Israeli transport planes to return empty. This note from the Den Uyl archive may well indicate that the premier clearly knew what was going on.

Meanwhile, Ambassador Bar On had convinced Vredeling, Van der Stoel as well as Den Uyl of Israel's great need of ammunition. This request set a series of activities in motion. Vredeling asked the Secretary-General of Defence, G.H.J. Peijnenburg, to obtain information from the Israeli military mission in Paris regarding Israeli wishes. It was in the meantime known that there was a special need for 105 and 155 mm artillery. It was not only Vredeling who was trying to clarify the situation. Van der Stoel instructed his Ambassador in Washington, R.B. van Lynden, to find out what the US Government thought The Netherlands' contribution should be. And Den Uyl requested Kruimink to draw up a memorandum over arms deliveries.[45]

The information requested by the Ministers was presented the following day. Vredeling received a memorandum from Peijnenburg, whose information had been obtained by the Quartermaster General, J.L. Antonissen, who in turn had been instructed by the Israeli military attaché in Paris. The Israeli reply was clear: Israel needed as much as possible 105 and 155 mm ammunition of any type, both for cannons and howitzers. Antonissen informed Peijnenburg that The Netherlands had no surplus stocks of this ammunition, i.e. stocks beyond those needed for exercises and in case of war. Stocks of 105 mm in particular were still being built up.[46]

According to Peijnenburg, The Netherlands could nevertheless relinquish munitions and reorder replacements from *Eurometaal* (the former Artillery Establishment). Delivery of 155 mm shells was easier than the 105 mm ammunition, because the English would be able to cite licence restrictions as an objection to their being re-exported to Israel. Antonissen thought this unlikely, however. Peijnenburg, concluding his advice, stressed that the Cabinet, or a few Ministers, must now decide whether deliveries to Israel should be permitted to go ahead. He pointed out that in 1967 deliveries had been made out of Dutch stocks and that this had been successfully carried out without publicity. Peijnenburg had meanwhile let

Bar On know by telephone that the Israeli military attaché in Paris had been contacted to establish exactly what Israel's needs were, and that once this was known, the decision lay with the Cabinet. Bar On was told no more than this by him.[47]

More information arrived at the Ministry for Foreign Affairs on October 9. From Washington came the news that The Netherlands must see what they themselves could do,[48] which meant that in any case there would be no American repudiation. Van der Stoel was kept informed of activities at the Defence Ministry. A Foreign Ministry memo of October 9 reported the Israeli request for 'any type and any quantity' of 105 and 155 mm ammunition. Defence had let it be known that smaller quantities of the above calibre were available from surplus stock and could in addition be supplied from stocks intended for the Dutch army's own use.[49]

The following morning, October 10, Den Uyl received a note from Kruimink titled: 'Several factors of relevance in evaluating the question: what is the value of 11,000 tank shells for Centurion tanks.' The note made reference to the possible delivery of 11,000 tank shells, field telephone cable, tank parts and also mines. The artillery ammunition that had been so centrally significant in the Defence papers was not even mentioned in Kruimink's note. According to him, what was important at that time was above all spare parts and tank shells for the Israeli Centurions.[50] This shift was probably linked with developments in the war, for after the first few difficult days, the Israeli army had now gone on the offensive.

The Israeli interest in ammunition and spares for Centurion tanks was understandable. These tanks constituted about half the Israeli tank force, in a situation in which they were confronted on both fronts with superior numbers of Syrian and Egyptian tanks.[51] The British Centurion tank was also the standard tank in both Dutch and British armies. The British government, however, as already mentioned, had banned the export of weapons to both belligerent parties.[52] For the Dutch army, the 11,000 tank shells constituted 'ammunition for the first phase', predestined for five days of Dutch fighting in the event of war (mainly in Germany, it was hoped), whereas this was probably sufficient to see the Israelis through three days of battle.

Given the content of his note, Kruimink was assuming a secret operation. He indicates briefly how transport to Israel could be worked out practically and also made suggestions for camouflaging the necessary replenishment of stocks in The Netherlands, which would have to occur in consultation with Britain. It was therefore inevitable, thought Kruimink, that Britain would have to be approached over this matter of re-stocking. Contact would also have to be made with the Americans.[53]

The problems were resolved on Wednesday morning, October 10, in a conference at the Dutch Foreign Ministry involving Den Uyl, Van der Stoel, Vredeling and the Director-General of Political Affairs, Van Lynden.[54] The position of the Foreign Affairs chiefs was immediately clear, both the Head of the Department for Africa and the Middle East (DAM) and Van Lynden opposed it. The head of DAM argued: 'Unless M (the Minister) definitively decides that under the present battle circumstances Israel is to be supplied with military materials, I would ask you to consider whether in this conflict situation any material should be supplied to any warring party that might contribute to the continuation of war.' Van Lynden also advised against supplying Israel unless Israel's own territory was under threat. In the course of these deliberations, Van Lynden referred the three Ministers to existing policy: 'not to supply the belligerent parties' and 'not to supply Israel or the Arab states since the forming of the new Cabinet'. Further, Van Lynden stressed the danger of reprisals by the Arab countries when it came to oil.[55]

Finally, the three Ministers involved decided to withhold supplies 'provisionally'. At least, that is what was noted on the Foreign Ministry memorandum.[56] It was certainly not a definitive refusal; quite the contrary, it was a decision which, in view of the attitude and choice of words of those involved, still left everything open. Vredeling, according to what he himself said, found the decision taken wholly unsatisfactory.[57] But Van der Stoel's subsequent account also leaves considerable room for interpretation. At the time, Dutch supplies seemed to Van der Stoel 'personally' unnecessary because American assistance with weaponry was already under consideration. Furthermore, as he explained some 25 years later, the ex-minister 'made a possible exception for *American armed materials that had been given us on loan'*.[58] This is a remarkable addition, since the urgent Israeli interest in Centurion spare parts did in fact concern material that The Netherlands had been 'loaned' by the USA in the mid-1950s, at the time of American military assistance under the *Mutual Defense Assistance Program (MDAP)*. Supplying Israel with Centurion parts and ammunition therefore, at least in part, did involve material given by the USA 'on loan'.

Van der Stoel also remembered that there had been talk of a swap, i.e. deliveries in exchange for later compensation.[59] In all probability this mooted exchange concerned the Centurion shells to be supplied. The fact is that stocks lay ready in Great Britain, destined for Israel, already paid for, but which because of the British arms embargo could not be delivered. The solution was simple: The Netherlands would supply Israel and would later get back these materials from the stocks lying ready in Great Britain. In this way, the problem raised by Kruimink could be solved, namely, how to restore stocks to the same level without drawing attention to oneself. Through such a swap, the relevant British military authorities need never be informed of the Dutch deliveries. There is another reason for assuming that the decision-making of October 10 went further than a simple 'provisionally not'. Den Uyl later remembered, without actually giving the date as October 10, that it was agreed it would be 'a good idea to transport reserve ammunition from Volkel'. As he added to the Dutch historian Grünfeld, 'in fact it never came to that'.[60] But this added remark is not correct: it most certainly did come to that.

In view of the content of Kruimink's note and the debates about an exchange, the indicators all suggest a *secret* operation. The decision 'provisionally not to supply' can also be interpreted as a decision for the time being to *not officially* supply, i.e. not according to all the stipulations in force. Subsequently, the Foreign Ministry would always deny that they had been involved in the surrender of arms export permits. However, an official procedure authorised by different departments was out of the question given the wording of Kruimink's note and the decision-making of 10 October.

Finally, a last point. There was a second important decision taken at that meeting. Both Van der Stoel and Den Uyl remembered that it was decided to offer the freedom of Dutch airfields for any possible American-Israeli airlift. Den Uyl later said that 'from our side we then offered them the use of our airfields'. The airfield primarily in question was that of Soesterberg. In the end, the offer was not taken up because US transports were routed via the Azores.[61] However, there were Arab accusations that The Netherlands and Portugal were the only NATO partners prepared to collaborate in setting up an American-Israeli airlift.

Arms Deliveries

Later on the same October 10, Vredeling informed Under-Secretary of Defence Bram Stemerdink that he had shared in the decision to supply Israel with weapons. He told Stemerdink that he had that morning thoroughly discussed the question with Van der Stoel and Den Uyl.[62] Vredeling and Stemerdink subsequently always maintained that they personally took this decision; that otherwise the whole process would have taken far too long, not least because of the anticipated opposition of Van der Stoel. But in the light of what we have already seen above, it is very much to the point to ask whether this picture is an entirely accurate one. Stemerdink has further since declared that 'there was political consent to supply whatever was necessary'.[63]

Vredeling and Stemerdink came to the agreement that, if the whole matter were leaked, the latter would take responsibility and if worst came to worst he would resign to avoid bringing down the Den Uyl Cabinet in its infancy. They decided naturally to deny any knowledge of the operation. Next, Stemerdink contacted the Quartermaster General, Antonissen, who was to lead the whole operation. The Under-Secretary of Defence did not know that Antonissen had already been busy since Monday – or even Sunday – drawing up an inventory of what could be delivered to Israel.[64]

The question of arms deliveries preoccupied Van der Stoel, and in particular the 'swap' discussed on Wednesday morning. This is also rather remarkable in view of the fact that the Minister should not have been fully informed. Stemerdink recalled subsequently that he had a conversation with his fellow party member about this whole affair on Thursday, October 11. Van der Stoel then returned to the question of whether The Netherlands would in fact be able to replenish stocks discretely after the war, Stemerdink setting out the reasons why this was not an insuperable problem. The stocks intended for Israel and now lying ready in Great Britain would after all be shipped to Rotterdam and with a little sleight of hand they could be unloaded. The 'swap' need never come to light.[65] Dutch stocks would thus by the spring of 1974 be completely replenished. This happened with the assent of Stemerdink's counterpart, the British Secretary for Defence; for by March 1974 the Labour Party had been returned to power in Britain, whereas a Conservative Defence Secretary would have undoubtedly declined to cooperate.[66]

Matters were efficiently expedited. According to Vredeling, the whole operation of 1973 was conducted on a need-to-know basis. At the ministry, Quartermaster General Antonissen, of course, knew about the whole operation, as did Brigadier General T. Meines, the logistics deputy working for the Quartermaster General.[67] Besides Antonissen, the Deputy Quartermaster General, General Major T.A. van Zanten, also knew about the operation.[68]

Meines confirmed that it was mainly tank parts and ammunition that were delivered to Israel, chiefly major components such as tank engines and various small spare parts for Centurions. Tank shells needed for the Israeli tanks were also sent. This material was fetched from the depots in Soesterberg and Utrecht and taken to Gilze Rijen. Material from Germany was also sent on. The ex-Israeli Ambassador Bar On has also said that it was mainly a matter of artillery ammunition, tank shells and spare parts.[69] Those directly involved later reported to Peeters that the tank shells were taken from the arsenals of the First Army Corps, and that the Centurion spare parts mainly consisted of shock-absorbers, gun turrets, caterpillar tracks, gearboxes and engines. But according to Peeters, that was not all. Machine guns and later parts for light AMX-tanks were also flown to Israel, together with 0.40-canons, 22 mm ammunition for aircraft artillery and thousand-pound bombs. If it suited Israel better, these '1,000 lbs' were flown by the Dutch airforce, sometimes in F-27's, to the American base Ramstein in Germany and there loaded into Israeli aircraft.[70]

Meines points out that the Army was busy changing over to the West German Leopard tank. The Centurion material was thus becoming superfluous, and parts could readily be disposed of. The Centurions did not belong to the Dutch, they were on loan. This was also the case with the spare parts, although over time the Dutch army had also bought reserve parts themselves.[71] As far as the American-loaned material was concerned, Vredeling later emphasised that The Netherlands was not in a position to dispose freely of the relevant parts. Kruimink also accepts in retrospect that Antonissen maintained contact throughout the whole operation with a military attaché at the American embassy.[72] This was probably the MDAP attaché or a functionary of the Military Assistance Advisory Group. We have already seen that there was no need to anticipate any problems from the Americans. Besides, Vredeling points out that the involvement of American Centurion material served a kind of 'camouflage' function, for in case of discovery, it could always be maintained that the Centurions were being given back to the Americans.

The material was transported to Israel in unmarked Israeli Boeing 707's from the Gilze Rijen military airfield, and according to Stemerdink also from Soesterberg, Ypenburg and Valkenburg.[73] These Boeings, which belonged to the Israeli airforce and the Israeli airline El Al, were sprayed grey to make identification more difficult. It is also possible that other 707's were hired from European airline companies. Indeed, KLM was very soon accused by the Arab side of being involved. The Israeli airforce itself commanded only seven Boeing 707's.

Each of the 707's would have made an intermediate stop at the Belgian airfield Melsbroek. The transports took place at night, most probably commencing the night of October 12, and lasted several nights, possibly from October 12 to October 14, the day the American airlift openly went into operation. The military historian J. Schulten believes that the Israeli aircraft flew within NATO airspace via civil flight corridors, giving Schiphol as their destination. The military personnel involved in the transport were mainly cadets of the Royal Military Academy in Breda, who were told that these were unexpected night-time exercises.[74]

It is not easy to establish just how important the Dutch deliveries were for Israel. Twenty years after the event, Vredeling gave his own view in a rather emotional fashion, piling on the agony and insisting on the significance of the supplies as though to justify his own actions. They were weapons, he said, that had been 'begged and pleaded for'. It was a matter of sink or swim, and therefore, acting entirely in a personal capacity, he had decided to lend Israel a helping hand.[75] In Kruimink's note, however, it had already been decided on October 10 that the weapons would arrive – and according to him did arrive – too late at the front to affect any 'sink or swim' situation. The Co-ordinator of the Intelligence and Security Services furthermore opined that Israel's survival was no longer in question after the fourth day of the war. In view of this, he called the Dutch contribution 'valuable', 'more than a token gesture', 'but not decisive'.[76] Nevertheless, at the time, Kruimink found these comments no reason to speak out against the plans in his note. He pointed out that 'if the Egyptians succeed in keeping parts of the Sinai' the consequences would be seriously detrimental to Israel.[77] The Dutch transports thus may not have played so much an important role in defending the state of Israel against a threat to its survival, as Vredeling subsequently claimed, but they surely did help Israel in regaining the offensive. The Dutch government, or at least the ministers concerned, had already adopted the standpoint that a return to the *status quo ante* would be highly desirable for Israel, since it would allow Israel to enter peace negotiations from a position of strength. Albeit on a modest scale, through its supplies of military material The Netherlands contributed to the realisation of this goal while the war was still in progress.

Foreign Ministry Denial

In October 1973, Kruimink thought it would be impossible to keep the arms supplies a secret. The transport and loading would involve hundreds of military personnel. Peijnenburg was less pessimistic since previously, in 1967, the public had successfully been kept in the dark. And Peijnenburg was for a long time right. The arms deliveries of the October War did indeed escape public attention. When this became no longer feasible, first Stemerdink and subsequently Vredeling took personal responsibility. Den Uyl and Van der Stoel, let alone other members of the government, had known nothing.

However, different individuals involved in the affair are of the opinion that it is highly unlikely that Den Uyl and Van der Stoel did not know. That, too, is the verdict of ex-Ambassador Bar On. He recalls that Van der Stoel was indeed initially unresponsive in his assessment. That was shortly after the outbreak of the war, but when the situation became more serious for Israel, the government – i.e. Den Uyl, Van der Stoel and Vredeling – decided to look at the Dutch position again. An actual airlift, they decided, was not possible. 'But the Dutch government did agree to the possibility of Israel purchasing ammunition, particularly artillery ammunition and shells'. On his own admission, Bar On had constant contact over the affair with both Van der Stoel and Den Uyl as well as Vredeling.[78]

The Foreign Ministry and Van der Stoel furthermore played a remarkable role when PSP member of the Second Chamber F. Van der Spek began to make trouble. On October 26, Van der Spek tabled written questions on the matter of arms deliveries. According to him, on October 12, 13 and 14, unmarked B-747's had landed at Schiphol to refuel and to transport weapons to the Middle East. Van der Spek wanted to know whether great risks had been taken. Was it likewise the case that on October 19 military aircraft had landed on their way to the Middle East?

This was close to the truth, and the answer to these parliamentary questions caused a number of problems. In a memo of November 2, Van der Stoel was informed that the draft reply should already by the beginning of the week, and with the utmost haste, be submitted to the Dutch Minister of Economic Affairs, Lubbers, and the Minister of Defence. But co-ordination led to considerable delay. Vredeling was meanwhile in agreement. In Economic Affairs and in Defence it was felt that, because of the political nature of the questions, Van de Stoel should be the first to sign it. The Minister was also informed that the Ministry of Economic Affairs wanted it borne in mind when framing the answer that 'no licences for the export of arms had been issued'. Beside this, Defence had let it be known that the Gilze Rijen airfield had not been used as an intermediate stop for military aircraft on their way to the Middle East.[79]

In his reply to Van der Spek's questions, Van der Stoel did not simply declare that no arms had been supplied to Israel. That would have been a lie. He answered, also on behalf of Lubbers and Vredeling, that since the outbreak of the war 'no licences for arms exports to the Middle East had been issued'. That applied also to arms in transit. This was obviously a hypocritical answer: of course, no official export licence had been granted, since the entire operation was conducted in secrecy. The specific questions over Schiphol were answered with explicit denials.[80] The flights had in fact been from Gilze Rijen and possibly other airfields.

Van der Stoel thought the formulation of the answers to Van der Spek's questions, 'no export licences', was probably chosen 'with reference to the loaned material made available by America'.[81] This is a remarkable comment, since the Dutch government had no authority to dispose of material given by the US on loan.[82] In retrospect, Van der Stoel also insisted on the formulation 'to have been able to say in all conscience that no licence was given for the export of weapons during the Yom Kippur War'. Neither he nor Den Uyl had been informed of arms deliveries.[83] This is, in the strictest sense, true. Bar On stressed in this connection that there had been no question of delivering *weapons*, but the supply of *ammunition and spare parts*.[84] Others involved, both at the time and later, may well have relied on this tactical but dubious distinction, but if aircraft machine guns and mines were also supplied, the distinction is invalid.

It is also not very plausible that Den Uyl was not immediately, or at least within a few days, fully informed. Den Uyl was certainly present at the discussions of October 10. Furthermore, he would in all probability have been informed of the transports soon enough by one of the intelligence services. After all, it was an operation in which, quite apart from the unrecognisable foreign aircraft, several hundred Dutch military personnel had been involved. Former member of parliament H. van den Bergh later brought to the world's attention the story that in December 1973 Israeli Premier Golda Meir had effusively thanked Den Uyl at the Socialist International for his support, and that Den Uyl was highly surprised at this expression of gratitude.[85] But was Den Uyl surprised because he knew nothing about it (as Van den Bergh suggested) or because Meir thanked him so openly in the proximity of other witnesses? Brandt was also warmly thanked by Meir, but the reason for that show of gratitude was evident, for it was well-known that the American war materials had also been flown from West Germany.

And indeed, Den Uyl need not have been so surprised. A week after the war broke out, the premier had received a cordial letter, dated October 10, from the Israeli government warmly thanking him for the Dutch support that had been highly important to Israel.[86] Furthermore, it would be highly unlike Den Uyl to be left uncertain. Rumours were already circulating in October over Dutch arms deliveries. According to Vredeling, Den Uyl never once asked him what was going on, which in the light of Den Uyl's curiosity, remarked on by Vredeling and others, could mean nothing other than that he was already fully informed. Bar On also claims that Den Uyl had been informed.[87] Ex-Minister of Economic Affairs Lubbers is of the opinion that Den Uyl had 'some knowledge' of the affair and that he suspected that Lubbers also knew.[88] Kruimink similarly thinks it highly improbable that Den Uyl knew nothing.[89]

These rumours were not only circulating in The Netherlands, but more significantly abroad. This was quickly evident from an undated memorandum to Van der Stoel, most probably written during the first days of the war. A report in the Dutch Foreign Ministry, most probably from a friendly intelligence service, reads: 'The Israelis are anticipating some ammunition shortage for their artillery and have requested 105 and 155 mm ammunition from The Netherlands, according to some Western Ambassador.'[90] In Washington, various people were told both of the Dutch willingness to make an airfield available and of the arms deliveries. When Ambassador Van Lynden held a conversation with the Deputy Secretary of State, Kenneth Rush, on October 30, the latter expressed his appreciation of the Dutch role during the war. Rush stressed how disappointing the attitude of the other European partners and Spain had been. He admitted that the member states of Europe had not always been adequately consulted, but it was intolerable that they had denied Americans the right to use their airspace or the facilities to refuel on European airfields or to move their own American materials. Some member states on which Washington most counted had let the USA down badly – meaning, of course, the UK. The Netherlands, Rush emphasised, absolutely did not belong to this category.[91] Four days earlier, on October 26, the American Ambassador had communicated to Van der Stoel the appreciation of his government for the Dutch attitude during the October war.[92]

An American official would later write in the *New York Times* that 'the Europeans, with the exception of Portugal and The Netherlands, had refused to have anything to do with US effort to resupply Israel with weapons, in some cases denying them overflight and refueling by American planes'. But this open reference did not go down well with the Dutch

Foreign Ministry; for the whole aim was that the Dutch attitude should remain secret. On November 6, in a request that he should inquire into various matters of world affairs, the Ambassador in Washington was asked to advise the Minister how 'to dispel the wholly *incorrect* impression that we allowed overflights and refuelling'.[93]

Vredeling recalls that James Schlesinger, American Defense Secretary, also knew of the Dutch arms deliveries. This was evident in December 1973, when Vredeling spoke with him in The Hague. Schlesinger was fulsome in his praise of The Netherlands. The American had learned the first line of the Dutch national anthem *Wilhelmus* by heart.[94] Van der Stoel also remembers that Schlesinger had thanked him for Dutch support 'with tears in his eyes'.[95] If Van der Stoel really did think that no weapons had been supplied, he must have wondered what on earth Schlesinger was thanking him for.[96]

In October, it turned out that the French government had suspected something. Why, they wondered aloud in Paris, had The Netherlands not announced an arms embargo? The Dutch Ambassador in Paris was asked by The Hague to explain that The Netherlands had indeed announced no arms embargo but that the delivery of war materials was subject to approval. The Ambassador was to declare the following:

> In accordance with the principle always adopted that no supplies will be delivered to the belligerent parties, no supplies have been delivered to the countries concerned since the outbreak of hostilities.[97]

In early November, it appeared that London was also giving credence to the rumours of Dutch arms deliveries. After Ambassador W.J.G. Gevers had reported this, Van der Stoel instructed him on November 8 to take steps, employing the cryptic denials he himself had used in his answer to the questions put by Van der Spek.[98] This Gevers did, and the Foreign Ministry appeared to regret that the impression had arisen that the British government gave any credence to suggestions of Dutch arms deliveries.[99] But the British apparently did not entirely trust this, for on November 10 a British diplomat asked whether The Netherlands had imposed an embargo on the export of arms and ammunition to the warring countries.[100]

As Van Lynden, the Director-General for Political Affairs, had stressed during the deliberations of October 10, risks had indeed been taken, particularly if Arab countries got wind of the secret arms supplies to Israel. The Saudi Oil Minister Yamani later said that he knew what had gone on and that the arms supplies were the direct reason for the oil embargo being imposed against The Netherlands.[101] It is possible that Yamani was bluffing, but it is certainly true that some Arab countries accused The Netherlands at the time of giving military support to Israel. On October 16, the Dutch Ambassador in Damascus, J. van Hoeve, reported that the Syrian authorities were accusing the Royal Dutch Airlines, KLM, of being involved in the transport of weapons to Israel. He was summoned for a hearing with Vice-Minister A. Rafai, who 'wanted to draw attention to the fact that, according to Syrian intelligence, fifteen aircraft loaded with weapons and ammunition had departed from Schiphol'.[102] A day later, Van Hoeve stated specifically that it was supposed to have been *Galaxy* aircraft involved. This could only refer to the American airforce. The Dutch Foreign Ministry informed their Ambassador that the Minister had learned with displeasure 'of continuing Syrian accusations against The Netherlands on the basis of completely unfounded rumours concerning arms supplies to Israel'. Van der Stoel was also to make his displeasure clear to the Syrian Ambassador.[103]

There were more indications that Arab states suspected something. During a conversation on October 17 with four Arab Ambassadors, according to the report of that meeting, Van der Stoel had to defend himself against accusations of arms transports.[104] In fact, it is rather remarkable that the Arab countries have never publicly accused The Netherlands of arms deliveries. There were, however, sufficient other grounds for accusing The Netherlands of adopting a pro-Israeli attitude.

European Political Cooperation

On October 12, the war issue was for the first time raised in the Dutch Council of Ministers. Meanwhile, the Dutch position was becoming more difficult. KLM was already openly suspected of having transported persons and materials. During this Cabinet session the question of possible Arab oil sanctions was aired for speculation. Van der Stoel still had no indication that the oil supply was to be restricted, but assumed that the risk of this happening was very real if the war were to escalate.

On the same day, in Copenhagen consultations were being held within the *Comité Politique* of the Director-Generals for Political Affairs (DGPA) of the nine EC countries. The DGPA's considered the question of whether a declaration by the Foreign Ministers over the war was called for. This meeting took place within the framework of the European Political Cooperation (EPC) gathering of European Foreign Ministers.[105] A draft declaration was drawn up that was clearly based on the British and French positions. The declaration called for a cease-fire and for a political solution on the basis of the French interpretation of Security Council resolution 242, which demanded that Israel withdraw from *the* occupied territories. In addition, in the last sentence of the declaration the UK and France were given the mandate to represent the whole EC in the Security Council.

Van Lynden's initial instructions were to block any joint statement over the Middle East war in the *Comité Politique*. The instruction held that the war in the Middle East did not lend itself to decision-making, but only to consultation. The mandate granted to the UK and France to represent the Nine in the Security Council would simply mean that these two countries would put their own views into effect. The Dutch standpoint with regard to the cease-fire was also out of harmony with the draft text. The Hague wanted withdrawal behind the existing truce lines as a condition for a cease-fire to enable a formula acceptable to both sides to be reached. This was a further reason to withhold assent from a communal declaration.[106]

Van Lynden was absolutely opposed to this and further considered these instructions as having originated under the influence of his opponent, J. Meijer, Director-General for International Cooperation. Van Lynden feared, together with many officials in the Department, that The Netherlands was heading for isolation within the EC. Furthermore, blocking a communal EC standpoint would only succeed in weakening the West European position with regard to the Arab countries. Van Lynden later claimed that decision-making within the Foreign Ministry was influenced by a powerful pro-Israeli lobby, to wit from the Directorate General for International Cooperation.[107]

The Dutch standpoint indeed encountered serious objections in the *Comité Politique*, leading Van Lynden to seek contact with his Minister who, at the time, was in ministerial council. Following Van Lynden's advice, Van der Stoel agreed to a communal EPC standpoint. Van der Stoel even instructed him to go along with an appeal for an immediate cease-fire, without further reference to any restoration of the status quo ante. It should be noted at this point that Israel's military situation had in the meantime improved considerably. Van der Stoel nevertheless continued to have great difficulty with the final passage of the proposed statement in which the nine member states empowered the UK and France to represent the EC in the Security Council.

During this Cabinet session, Van der Stoel informed his colleagues that he had, meanwhile, given instructions to strike out the final passage concerning the British-French mandate. He saw no way in which the Dutch and the French visions could be brought into harmony with each other with French representation in the Security Council. In general, any initiatives in New York should preferably be taken on the basis of the old truce lines. And, it should be said, Van der Stoel was certainly not alone. Transport Minister Westerterp pointed out that Paris had meanwhile put out a statement in which the Syrian and Egyptian attack had been more or less approved. He reminded his colleagues that France had also abused such a mandate in the past: one could well guess what would happen with a new mandate. Van der Stoel agreed wholeheartedly with this point of view.[108]

In the *Comité Politique*, in accordance with the instruction from his Minister, Van Lynden stated that The Netherlands found itself unable under any condition to agree to the last sentence of the draft declaration. As a result, the Dutch Director-General for Political Affairs found himself heavily attacked, particularly by the Italians. Italy and France refused to remove the last sentence, leading to an impasse which was only broken the following day when the French and Italian governments gave their consent to dropping the mandate. When the statement was published on October 13, most Arab countries reacted with disappointment or even anger. The demand was heard almost immediately that the EC should declare itself in favour of complete implementation of resolution 242, and specifically for complete withdrawal from *the* occupied territories.

A veto was thus declared, under the responsibility of Van der Stoel and with the support of the Cabinet, on the mandate requested by Paris and London. Without doubt, this had to do with the differences of opinion concerning the Middle East conflict. However, it should be noted that European Political Cooperation (EPC) in the field of foreign policy was still at a very early stage. This was one of the first times that the EPC had been considered as the appropriate, co-ordinating body to act in such an important matter. The Netherlands had long stood out against such political cooperation between the six, and subsequently the nine member states. The veto in part possibly stemmed from the usual aversion to the politicizing of European integration, although it should be added that there were fewer misgivings on this score within the Directorate-General for Political Affairs than entertained by the Minister and within the Cabinet.[109]

Like a true Atlanticist, Van der Stoel took for granted American leadership in the question of the Middle East conflict and advocated first of all Atlantic unity rather than political co-ordination within EPC.[110] This did not prevent considerable dissatisfaction in The Hague with the American government's policy, particularly over the lack of consultation with the

West European allies. When the new American Ambassador in The Hague came to pay his respects to the Secretary-General of the Dutch Foreign Ministry, E.L.C. Schiff, the latter made it very clear that The Netherlands also needed to be kept in touch with the thinking and actions of the American government. Understanding and cooperation could hardly be expected of the West European allies if they were not adequately informed of American objectives and tactics.[111]

On October 16, the American government declared the war in the Middle East a 'major test' for NATO. Unity was necessary in the face of the Soviet threat. Responding to this appeal for Atlantic solidarity, Van der Stoel expressed himself cautiously positive. In an instruction to Fack, the Dutch Minister stressed that, given the Soviet backing for the Arab countries, American military support for Israel was understandable. He had sympathy for the American resumption of arms deliveries and subscribed to the belief that the NATO allies must do everything possible to maintain their unity. He advocated considering measures against the Soviet Union, for example within the framework of CSCE consultations. Unless Moscow exercised the necessary restraint in time, thought Van der Stoel, the climate in Geneva would be badly affected.[112]

Nevertheless, within a few days, West European resentment was again aroused by American action when Nixon, as mentioned earlier, put his troops on a state of high alert. On October 26, the NATO Permanent Representatives met in Brussels. The West German government, meanwhile, had taken the decision to end its collaboration in the transport of American arms stocks to Israel. Although Israel had the winning hand, Bonn appeared to side with France and Great Britain on this point also. Both the French and Belgian Permanent Representatives sharply criticized the American conduct, specifically the lack of information given – particularly when the Americans had so shortly beforehand appealed for solidarity and cooperation. The Dutch Permanent Representative let it be known that, in spite of everything, The Netherlands could understand the American actions.[113]

Conclusion

The Dutch government's policy, as we said, did not go unnoticed in the Arab world. There were definite rumours of Dutch military assistance to Israel. Further, the Dutch veto of the EPC mandate requested by Paris and London did not long remain a secret. On the basis of information from 'highly qualified sources', *The Guardian* reported that 'France and Italy privately informed Arab governments about Dutch reluctance to support a joint neutralist European stand on the Middle East'.[114]

That was not all. On October 13 a large demonstration had been held in Amsterdam in support of Israel, which Defence Minister Vredeling attended. On October 17 there appeared a page-sized declaration of solidarity with Israel in several newspapers, signed by many prominent public figures from Dutch society, including ex-ministers and ex-premiers. The presence of Vredeling at the October 13th demonstration seemed to demonstrate clearly that, in this issue, the Cabinet stood squarely behind pro-Israeli public opinion. In the coming weeks, the Arab countries would return time and again to this remarkable gesture on the part of Vredeling.

Why did the Den Uyl Cabinet adopt such an emphatic position? The normal alliance with Israel is a possible explanation but certainly open to doubt, for in the preceding years there had been a certain degree of estrangement. The party political composition of the Den Uyl Cabinet played a part. Those members of the government most involved were the Dutch Labour Party (PvdA) members who (more than, for example, the Catholic People's Party, the KVP, according to ex-minister Van der Klauw) felt solidarity with Israel and certainly also with the governing Israeli Labour Party at that time.[115]

The Dutch government rejected the understanding of the Arab countries advocated by the French, and resisted any joint EPC role, particularly in the Security Council. Indeed, Van der Stoel's preferred orientation was pro-US rather than pro-EPC, which conformed to the usual line of judgement where European Political Cooperation was concerned. Nevertheless, Van der Stoel's policy is remarkable, given the criticism voiced by the DGPA and by diplomats. In the weeks that followed, Van der Stoel's judgement was not always shared by his political and diplomatic advisors.

Furthermore, arms were delivered, even though the Cabinet had never taken any decision on this. Vredeling always insisted that he had acted on his own initiative, together with Stemerdink; but in light of the above, that hardly appears likely. It would seem more probable that the various individuals involved were fully informed or at the least had their suspicions. Vredeling subsequently confirmed that none of his colleagues ever asked him explicitly about the arms deliveries, even though various rumours, questions and accusations were flying around. When Stemerdink years later told Den Uyl what had happened, Den Uyl said that he always suspected something of the kind. It is quite extraordinary that Den Uyl had never asked his Defence Minister for clarification. The inescapable conclusion is that the premier knew very well what was going on.

The attitude taken by the Dutch Cabinet carried necessary risks, in the first place the risk of Arab sanctions. In the coming weeks, Van der Stoel would be confronted with Arab accusations on various occasions. However, it is remarkable that when Algeria pronounced an oil embargo against The Netherlands (the first country to do so), although a series of accusations were thrown at The Hague, these did not include any accusation of arms deliveries.

The Cabinet position also carried with it problems within the EC; though at the same time the support of Washington could be relied on. Indeed, the Dutch Ambassador in Washington, Van Lynden, was told on October 30 just how highly the Americans appreciated the Dutch role. Besides, in the event of conflict in the oil sector, The Netherlands was in itself in no weak position:[116] the home base of Shell, it included Rotterdam and the huge refinery capacity in the Rijnmond area. Furthermore, it commanded considerable reserves of natural gas which it exported to Belgium, West Germany, Italy and France. This fact may well have played a role in determining policy with regard to the October war and the subsequent oil crisis.

All in all, risks were taken, without a doubt, and yet at the same time the Dutch government was sailing a course in the wake of the USA with a certain accuracy. This is not to deny that there was sometimes irritation in The Hague at the lack of information provided from the American side. Van der Stoel complained of this in a conversation with the American Ambassador.[117] Yet perhaps there was at the back of the minds of all the members of the Dutch government the conviction that if it came to the crunch, the Americans would not abandon their small but steadfast ally. There were times, as Van der Stoel put it, when solidarity between NATO allies was essential.

2
Difficulties

In the years leading up to the 1973 war, the international oil sector had undergone structural shifts against the background of a world-wide increase in the demand for oil.[1] Oil production had also increased hugely, not least in the Arab countries. Between 1950 and 1973, oil extraction in the Middle East increased fourteen-fold.[2] Western Europe and Japan in particular had become increasingly dependent on Arab oil.

The West European countries had become to a significant degree dependent on oil imports for both their energy production and their petrochemical industries. In 1955, coal was still the most important energy source in Western Europe (75%), whereas in 1972 this share had fallen to 23%. In contrast, oil had risen from 22% to 60% of the total energy supply. In 1955, 78% of all West European energy needs were met from within Western Europe, whereas in 1972 this figure had been drastically reduced to 35%. Energy imports had correspondingly risen from 22% to 65%.[3]

The dependence on Arab oil imports, as a percentage of total energy requirements, rose in Western Europe from 13.4% in 1956 to 36% in 1967 and 45% in 1973. This in turn had considerable consequences for the status of the oil-producing countries of the Middle East, particularly Saudi Arabia, whose share in the world's oil production rose from 13% in 1970 to 21% in 1973. This development made Saudi Arabia one of the most important political players in the Middle East.[4]

Even in the USA, long self-sufficient in oil, scarcity began to make itself felt on the oil market. The oil-producing areas in the USA could no longer satisfy the ever-increasing demand, with the result that oil imports grew throughout the 1960s at an average rate of 6% – rising to some 20% after 1970. The squandering of reserves and increasing consumption meant that the USA had become a net importer of oil. Although American dependence on imports was far less than in most other industrial countries, the term energy crisis became securely established in the vocabulary of American politics.[5]

Because the countries of the industrial West were more dependent, they became more vulnerable. Furthermore, the powerful position of the traditional, international oil companies, the *majors* or *Seven Sisters*, symbols of Western domination of the world economy, appeared to be under pressure by the early 1970s. In 1972, although the *Sisters*, consisting of five American, one British (BP) and one Anglo-Dutch concern (Shell), still controlled 71% of oil production (excluding the USA and the communist countries), 49% of refining capacity and 54% of product-marketing, the power of these companies was on the wane. The ambitions of the oil producing countries were clearly spelled out by nationalisation in Algeria and Libya, in 1970 and 1971; ambitions which were expressed in the activities of the *Organisation of Petroleum Exporting Countries* (OPEC) which had been set up in 1960. The status of the *majors* was further seriously affected over the years by the so-called independent American oil companies (Occidental, Phillips, Getty, Standard Oil of Indiana), as well as by the rise of national and semi-national concerns such as the French CFP and Ef-Erap and the Italian ENI and Agip. The emergence of these new competitors appeared to offer the oil-producing countries new perspectives. The new rivals needed to strengthen their position and were therefore likely to show a greater willingness to meet their demands than the *majors*.[6]

The growth of independent oil concerns and of West European and Arab national oil companies was not, however, to the advantage of the oil producers in every respect. It also meant that ever more oil was coming on the market, with inevitable effects on oil prices which, initially as a result of the enormous supply in the late 1960s, gradually declined. From that moment, the OPEC countries, with considerable verve, tried to push through a new policy. The number of OPEC member countries had meanwhile sharply increased from five founder members in 1960 to thirteen countries in 1973. In particular, the entry of Algeria (1969) and the radicalisation of Libya led to an ever greater pressure on traditionally conservative oil producers like Saudi Arabia.[7] OPEC began to force up oil prices. This was understandable, since for decades oil prices had hardly increased at all, while the prices of Western oil products had often increased by dozens of percentages if not a hundred percent. OPEC also began to insist on a greater say – or 'participation' – in national oil extraction. Further, the OPEC countries wanted to manage their mineral resources intelligently and, if necessary, to limit oil production. After all, a reduced supply should raise the price of oil.[8]

In December 1970, by which point oil prices had fallen to an all-time low, the OPEC countries demanded steep rises both in the price of oil and in taxes on oil production. At first, the *majors* declined to cooperate with this demand; while for the first time ever, the so-called 'parent countries' of the main oil companies – among them The Netherlands, albeit rather reluctantly – met to discuss this development in Washington. On February 14, 1971, agreement was reached on the OPEC demands when the main companies and six Arabian Gulf states concluded an agreement that included increases in oil prices and taxes. There subsequently followed similar agreements with other oil-producing countries. A few Arab OPEC countries, such as Libya, were not happy with this February outcome, and negotiations were reopened. On April 2, 1971, the Tripoli agreement was signed, setting a further increase in the price of oil.

Western countries' fears that these concessions would not be the end of the matter were very quickly realized. Within months the OPEC countries were demanding a greater participation in national oil extraction (mainly controlled by the *majors*) as well as compensation for the devaluation of the dollar, the currency of payments for oil.[9] In January 1972 these aspirations led to a new agreement between the Gulf States and the companies on oil revenues and to the setting up of discussions on expanding national participation which, if OPEC had its way, allowed member states a 51% national share.

The apparently insuperable clash of interests between OPEC and the oil companies began to look as though it were not, after all, entirely irreconcilable, even if certain aspects could not be resolved. The *majors* were thus not dissatisfied with the moderate price increases agreed in 1971 and 1972. In reaching these settlements, the *majors* were helped by Saudi Arabia, which feared that too steep a rise in prices could perhaps lead to an uncontrollable inflation in the industrialized world, in which case Saudi investments abroad would be severely devalued. There was the further fear that the search for alternative energy sources would be stepped up in earnest.

Gradual price rises suited the long-term strategy of the *majors*. Moderate increases coupled with a guaranteed supply could well deliver greater profits which, at the time, the *majors* certainly needed. Meanwhile, they were on the lookout for new oil areas in politically safe regions beyond OPEC's demesne. They were also considering bringing into production such new oilfields as Alaska and the North Sea, more expensive locations that would call for extra capital if they were to be developed. Traditionally, the oil industry had always been able to finance its own development, but profits were no longer high enough for such large-scale operations. Deep-sea drilling in particular entailed colossal initial outlay, costs that could only be recouped by raising the price of oil.[10]

Nor was it only the *majors* who had an interest in higher oil prices. The American government, unlike the *majors*, had from the outset expressed no objection to higher oil prices, and in 1972/73 had informed OPEC accordingly.[11] Such increases would ensure that the profit margins of domestic oil companies in the US would rise, making it a more attractive proposition to utilize otherwise uneconomic oil wells in Texas and elsewhere. It should not be forgotten in this connection that in the early 1970s the US was dependent on foreign oil for a mere 15% of its total energy consumption. In Western Europe the situation was quite different. France depended on oil imports for 67% of its energy consumption, Italy for 74%, West Germany for 55%.[12]

Another effect of the hike in oil prices was that Western client states such as Iran and Saudi Arabia would acquire greater financial scope for large-scale technological and military purchases. The American presidential advisory body, the Council on International Economic Policy, anticipated that Saudi Arabia's oil revenues would increase from 5 billion to 20 billion dollars. In the mid-1970s the arms sales to Iran, Saudi Arabia and Israel amounted to more than 50% of all foreign American arms sales. Besides which, the State Department's view of the oil price increases could only benefit American business life, since such a blow was mainly felt by America's European and Japanese competitors, whose products would become more expensive. Washington's stance, therefore, was initially not unsympathetic toward OPEC's demands.[13]

However, the scenario did not unfold entirely as anticipated. The OPEC countries turned out to be unwilling to go along with gradual price increases, while mutual tensions between the consumer countries generated ever more problems, as became all too apparent at the time of the oil crisis. The oil market was becoming increasingly politicised. One of the first reasons for this was that the governments of the consumer countries, the *parent countries*, found themselves more and more embroiled in questions of oil. Against this background, and to an increasing extent, differences of viewpoint emerged between the consumer countries – and not least between the Europeans and the USA. The position of the European countries vis-à-vis the USA was even more problematic as they could not succeed in speaking with a single voice. In the years leading up to 1973 there had been attempts within the EC to arrive at a common energy policy, but the clash of interests between member states had always stood in the way of such agreement.

Oil was not the only problem aggravating relations between the USA and the European countries in this period. In fact, the oil problem cannot be seen in isolation from other politico-economic conflicts. In 1971, not only had America's balance of payments been in deficit, but also the American balance of trade. Particularly in sectors like the motor industry, steel and electronics, European and Japanese producers had undermined the American position. Nixon's reaction was to cancel the dollar's convertibility to gold and to devalue the dollar, especially with a view to making exports more attractive.[14] This suggested that Washington was prepared to set off a trade war with Western Europe and Japan. Raising oil prices was a part of that trade war.

The politicising of the oil market was also a consequence of the increasing self-awareness evident in the actions of the oil-producing countries who were determined to get a firmer grip on their national oil production. In the Western consumer countries, such greater participation evoked only alarm. This fear, however, was not enough to prevent an accord being signed between the major oil producers, such as Saudi Arabia and Kuwait, who wanted a step by step increase in the national share in the oil production up to a maximum of 51%. For the *majors*, in retrospect, the consequences of such nationalization subsequently turned out to exceed expectations at the time, for this did not affect their position in the processing and distribution sectors at all.

In early 1972 the American State Department produced a secret report of some seventy pages on *The Impending Oil Crisis*, setting out the typical position in which the Western countries now found themselves, the USA in particular. The report assumed that around 1975, possibly earlier, the era of a permanent sellers' market would have arrived, with any of the several major producers being able to create a supply crisis by cutting off oil supplies. It was also assumed in this report that America's energy position would have weakened by around 1980 to the point where the USA would be forced to import more than 50% of its oil requirements.[15]

The consumer countries had survived the price increases carried through by OPEC in 1971, the report declared, as a result of mutual solidarity, but this success had been underestimated. In the view of the State Department, OPEC members must be convinced that long-term stability would avail them far more than complete chaos and short-term gains. It was recognized that the devaluation of the dollar had caused problems, but these need not be insuperable. The problem of participation was much more difficult, however, because the differences on this point did seem to be irreconcilable. The danger of nationalization could only be prevented in the long term if a serious dialogue were initiated to discuss a new relationship between the *majors* and the governments of the oil-producing countries. In addition to which, according to the State Department report, the American government had to consider a possible reduction in the growth of oil consumption, an increase in domestic production and importing from safer sources.[16]

The question of price increases meanwhile dragged on. In June 1973, a new increase of 12% was agreed, but this was still not enough. Just before the war broke out, new negotiations were announced between OPEC and the major oil companies, negotiations which should have opened on October 8 in Vienna. The OPEC countries wanted a 100% increase in the price of oil. This was no longer a matter of gradual price increases. Shell and Exxon, wanting first to consult with Western governments over such drastic steps, insisted on postponing negotiations, since the kind of increases OPEC wanted would have far-reaching consequences for the economies of the West. The negotiators, George Piercy (Exxon) and Andre Bernard (Shell), therefore decided to insist to the Saudi Oil Minister Yamani that the negotiations be postponed for two weeks.[17]

On the eve of the October War, there were thus a number of unresolved differences within the international oil industry brewing, and concomitantly, the developing fear in the Western consumer countries that the 'oil weapon' would be used politically. This fear was mainly focused on the Arab OPEC countries who since 1968 had amalgamated in OAPEC. In the early 1970s, under the influence of the more radical members, OAPEC announced on several occasions that oil wealth would be used as a weapon against Western countries that supported Israel, warnings which were to be repeated before the outbreak of the October War. On May 15 1973, the day Israel celebrated its 25 years of existence, Iraq, Kuwait and Algeria turned off the oil tap for an hour, Libya for the entire day. Nor did the radical oil producers speak for themselves alone. In the summer, the top executives of Aramco (the daughter company of Chevron, Exxon, Texaco and Mobil) held a meeting in Geneva with King Feisal of Saudi Arabia at which the latter threatened that, if America did not change its pro-Israeli attitude, Aramco's access to the Arabian oilfields would be withdrawn.[18] A resolution passed in the Kuwaiti parliament in June 1973 in the clearest terms called for a freeze on oil production in the event that war should break out in the Middle East.

Enough warnings had thus been sounded, and if that were not enough, the Libyan leader, Colonel Qaddafi, declared on July 10 on French TV that in future the Arabs would be able to use their oil as a political weapon against the USA and Western Europe. On September 4, too, at a meeting in Kuwait of the Foreign Ministers of the OPEC countries, the question of using the oil weapon was openly discussed. The claim made by Yergin, in his book *The Prize*, that the embargo came almost as a total surprise cannot therefore be taken seriously.[19] The use of the oil weapon most certainly did not come out of the blue.[20]

Turf War in The Hague

During the years leading up to 1973, the possibility of restrictions on the oil supply had also preoccupied The Hague. The Dutch Ministries of Economic Affairs and Foreign Affairs in particular had been at work on this question: indeed, there had been a certain vying for recognition between these departments as to whose area of competence it was. Economic Affairs was responsible for foreign economic relations, which was taken to include the question of oil supplies. Yet to the extent that this question of oil supply was being raised increasingly often in such international organizations as the Organization for Economic Cooperation and Development (OECD), and the more the political nature of the whole oil question was foregrounded, the more the Foreign Ministry became involved. During the period of the Biesheuvel Cabinet (1971-1973), this led to friction between the ministers concerned, H. Langman, Minister of Economic Affairs, and W.K.N. Schmelzer, the Foreign Minister.

At the Dutch Foreign Ministry, the oil problem was especially the concern of the Department for Economic Cooperation (DES). The central role of DES stemmed from the increasing importance of the OECD and the EC in matters of oil. In January 1971 it was decided to appoint the DES as co-ordinating organ within the Foreign Ministry where matters of oil supply were concerned. Plans for emergency allocation had been debated within OECD for some years. At the beginning of 1970, prompted by the Americans, the OECD again began to pay more attention to the security aspect of the oil supply. Washington's basic position was that within a few years the West European countries could no longer expect to be able to call on American oil reserves. In addition, the demand for crude oil on the world market was now in excess of supply. The US therefore wanted the OECD to discuss measures for reducing dependence on Arab oil and, if necessary, together to prepare steps in the event of a reduction in the supply of oil.[21]

The OECD could not agree on the issue of an allocation system in crisis situations. There was no problem in reaching accord on the need for laying in emergency supplies, but the OECD was not in a position to force member states to act on this. The EC or the European Commission were however, and although it had not been possible within the EC to agree on a common energy policy, the European Commission issued a directive that all member states were required to lay in emergency supplies: for 65 days with effect from January 1, 1971, and for 90 days with effect from January 1, 1974.

The change of government in 1971 brought no change in the sharing of responsibilities between the Ministries for Economic Affairs and Foreign Affairs, although Minister Langman suspected his colleague Schmelzer of wanting to take over the primary responsibility for the oil issue: not without justification, it seemed. There was indeed a level of unrest within the Foreign Ministry over the dominant role played by Economic Affairs in the matter of oil supplies. On January 4, 1972, Schmelzer informed Langman that he considered himself responsible for oil politics in the international context; but this proposal led to no revision of the division of competencies between the two ministries. Schmelzer then proposed letting the question rest. DES subsequently tried to come to some arrangement with Economic Affairs separately over a ruling at the executive, but this offer was also declined by Economic Affairs.[22]

True to its traditions, The Netherlands had adopted a liberal attitude toward the international oil market and had spoken out against market regulations and state intervention. In 1971, the government had taken a reserved position in the consultations in Washington between the parent companies and the major oil companies. A report drafted in 1971 on the vulnerability of Western crude oil supplies reiterated this assessment. Intervention by the state would involve untold risks, not least because the OPEC countries would then be inclined to raise political issues. The *majors* would be in a better position to deal with OPEC than national governments, who were not only far more divided among themselves but also susceptible to the influence of domestic political interests.[23]

Yet the tenability of even this rather conservative standpoint was doubted. In 1972 the Head of DES concluded that it was inevitable that governments would be drawn into the problems of international oil, and that the government therefore had to develop a clear international energy policy. Whether within or outside the frame of the EC, the question that had to be investigated was whether improved relations with the oil countries would promote the secure future provision of oil.[24]

In the spring of 1973, the policy unit (PLAN) of the Foreign Ministry was asked to draw up a memorandum on this question. Subsequently and after extensive preliminary discussions, a couple of papers were submitted by H.Ch. Posthumus Meyes, advisor to the PLAN Policy Unit, to Foreign Minister Van der Stoel in May 1973: *The oil crisis and Dutch foreign policy* and *The use of the oil weapon for political purposes.*[25] In these papers, Posthumus Meyes reached the conclusion that, where oil supplies were concerned, the 'careless period' of the past was over. Oil would become an increasingly important subject of government interference. Complete abstention by the state was no longer tenable, though governments should not assume the role of the oil companies at the negotiation table. More active involvement on the part of the consumer countries would also involve risks, such as rivalries in the 'scramble for oil' and the use of oil as a tool for applying political pressure. Care should also be taken to avoid a buyers' cartel being set up that might lead to a dangerous confrontation with the producer countries. Consultation between consumer countries should, among other things, look at controlling competition, promoting the diversification of supply and energy, emergency provisions and activities to channel the enormous financial resources of the producer countries advantageously. Although The Netherlands, according to Posthumus Meyes, was well situated, given its North Sea natural gas fields, its major oil industry and the position of Rotterdam, it was nevertheless too weak to pursue an independent course in matters of oil.

In *The use of the oil weapon for political purposes*, Posthumus Meyes particularly pointed out the possibility of oil being deployed as a weapon in any Middle Eastern conflict. During such a conflict, the threat to the oil supply could be exerted to force the West to take a different attitude toward Israel. There was also the possibility that dependence on Arab oil might be exploited to incite Western countries into military support against the Arab countries. In any such situation, the memorandum proposed, '*the Netherlands would also be involved*'. The role of the USA in the Middle East was still dominant. The question that arose was whether Western Europe should not take its own, independent line, such as up to that point had not been possible. It was a better idea to lend support to 'any American turn toward a rather more balanced treatment of the Middle East question.'[26]

PLAN's memoranda were sent to various Ambassadors with the request for feedback, provoking a reassuring response from several quarters. W.J.G. Gevers, the Ambasssador in London, opined that in the short term there seemed to be no major problem likely to arise, and that in any case the Western countries, including The Netherlands, held stocks sufficient for ninety days. H.N. Boon in Rome likewise thought an embargo unlikely, although restrictions on production should not be ruled out.[27] From Washington, D.W. van Lynden reported that the State Department, albeit in the view of Under-Secretary Kenneth Rush, considered the political use of Arab oil unlikely, particularly in view of the divisions extant in the Arab world.[28]

However, The Netherlands' Permanent Representative (PR) at the OECD, J. Kaufmann, found that the PLAN memorandum was too reserved in its appraisal of international cooperation, and its proposals for future possible influence by The Hague too modest and self-effacing. Because of the oil companies established in The Netherlands and other sectors of business life contributing significantly to the know-how and the provision of materials for oilexploitation, he thought The Hague should have an important voice in consultations among consumer countries.[29] Ambassadors in some of the Arab countries were also asking for a more active role. Cultivating extra goodwill in these countries was a real possibility. Saudi Arabia, for example, was attempting to broaden its economic base, and Saudis therefore were highly interested in joint ventures with foreign contractors.[30]

Nor was it only in the Dutch Foreign Ministry that serious thought was being given to the possibility of restrictions on the oil supply. After an extensive exchange of ideas, a crisis scenario was drafted in the Ministry for Economic Affairs. Distribution plans were laid, although these were to play no further role in the weeks ahead.[31] In various places, thus, the possibility of a restriction on oil imports was being given serious attention. And yet, in the spring of 1973, The Hague appears to have been totally unaware of one acute danger, let alone the fact that Arab measures directed specifically against The Netherlands were at that very moment being considered.

The First Signs

Shortly before the outbreak of the October War, as we have already indicated, disturbing signals were received. Rumours concerning Arab oil measures were already circulating by early September. In addition, Libya had decided to press ahead with its nationalization of 51% of all foreign oil interests. One of the places where this was discussed was the EC, in the EPC Middle Eastern working group as early as September 3. This discussion revealed differences of judgement and a remarkable divergence of viewpoints. On the one hand, Italy and France, countries with large state oil companies, thought that oil would play a political role within a very short time and that the EC would have to adjust to this. The traditionally liberal parent countries of Shell and BP, Great Britain and The Netherlands, however, did not share this point of view.

During the first week of the war, it appeared that the threat of an embargo would not materialize despite an appeal for such a measure from the PLO leadership. Lubbers and Van der Stoel confirmed in the Dutch Council of Ministers of October 12 that so far no political oil measures had been implemented, though the possibility remained. Even the radical Libyans were delivering normally. According to Lubbers, the central question determining the way the OPEC countries behaved was still the question of prices. After all, far-reaching demands from the Arab countries remained on the table at the Vienna negotiations which, in the meantime, had been interrupted by the war.[32]

As the war entered its second week, accusations began to be heard from the Arab capitals concerning Dutch political and military support for Israel. Syria in particular accused The Hague of military activities on Israel's behalf. Further, on October 16, the Arab OPEC countries met in Kuwait where they were to take a series of drastic measures whose consequences, it was announced, would cost countries that supported Israel dearly. It was this announcement that triggered a much-needed sense of urgency in The Hague. This appeared to be for real.

Van der Stoel himself was also now confronted with the Arab accusations. On October 17, at their request, the Minister received the Ambassadors of Iraq, Saudi Arabia and Tunisia together with the Egyptian temporary *chargé d'affaires,* who claimed to speak for the other Arab countries recognized by The Netherlands. At this meeting, the four diplomats appealed to the EC countries for some contribution to the resolution of the Middle Eastern conflict. In response to their plea, Van der Stoel put the case that the Nine could do little more than call on the warring parties to cease hostilities in accordance with the EC declaration of October 13. Concluding this conversation, according to the Dutch Foreign Ministry record, he informed them that he was not favourably impressed by exaggerated official reactions from Arab capitals to various rumours of volunteers being recruited and transported together with armaments from The Netherlands.

He requested that the four Ambassadors should again advise their governments of the baselessness of these reports.[33]

In retrospect, with the insight provided by the Foreign Intelligence Service (IDB), it would appear that, after their conversation with Van der Stoel, the Ambassadors suggested to their governments that The Netherlands should be warned against proceeding with political and military support for Israel; and that if the reaction to such a warning were negative, The Netherlands, like the USA, should receive no more oil. All Arab countries should be urged to condemn the clearly hostile attitude of The Hague. The Arab countries should furthermore blacklist all Dutch firms and threaten KLM – as well as other airlines – with reprisals if they continued to ferry mercenaries and arms to Israel.[34] These reports were to reach The Hague only later. For the time being, it was not clear what concrete sanctions the Arab side could impose. If it should come to that, according to Lubbers speaking in the Second Chamber on October 18, then the consequences of such sanctions would have to be taken care of through joint action in common with the EC.[35] A few weeks, however, would be sufficient to make this judgement appear naively optimistic.

The Oil Weapon Brought to Bear

The situation was gradually beginning to bite deeper. A number of important decisions were taken at the conference of the OAPEC held in Kuwait from October 16 until October 21. To begin with, in the first place by the six Gulf States, it was decided on 16 October to raise the price of crude oil unilaterally by 70%, thus by-passing the Vienna negotiations suspended on October 9. The oil companies would have to pay henceforth not $1.80 but $3.06 a barrel in taxes and royalties. In the second place, it was agreed to use the Western dependence on oil as a political weapon in the war with Israel. Saudi Oil Minister Yamani informed the oil companies on October 16 that his country would be cutting oil production by 10% if the US continued to replenish Israeli military losses. The following day, the eleven OAPEC countries decided to reduce oil production by 5% each month until Israel had withdrawn from all the occupied territories and had recognized the rights of the Palestinians. At the same time it was announced that those consumer countries that supported Israel would have to bear the consequences of these measures while those consumer countries that supported the Arab cause would be exempted from production restrictions.[36] As Yamani later commented, the Arabs thus finally showed themselves 'to be the masters of their own oil'.[37]

The American response to the Saudi threat was crystal clear. On October 19 Nixon requested congressional approval for 2.2 billion dollars in emergency aid for Israel. The Arab world reacted immediately: on the very same day Libya announced an oil embargo against the USA in response to arms deliveries to Israel. The following day Saudi Arabia followed suit, and on October 21 Kuwait, Qatar, Bahrein and Dubai joined them.[38] Panic ensued in the USA, for the embargo meant that some 3.5 million barrels a day would be lost. A Federal Energy Office (FEO) was set up with broad powers to compel reductions in the use of oil.[39]

In the Dutch Council of Ministers of October 19, the situation in the Middle East was again raised for comprehensive discussion. Meanwhile, the military situation in the region had drastically altered. According to Van der Stoel, the Israeli position had radically improved, particularly on the Egyptian front. The Syrian army had not yet been defeated, but there was now little chance of any new offensive against Israel. All in all, Israel appeared no longer to be in acute danger. Indeed, the territory now under Israeli control had expanded considerably compared with the old lines of demarcation.

Meanwhile, it had become clear that The Netherlands was likely to suffer the Arab sanctions. Van der Stoel recognized that relations with the Arab countries had been 'somewhat strained' following the Syrian accusation that The Netherlands had sent volunteer fighters to Israel. On various occasions he had protested about this, as in his meeting with the four Arab Ambassadors, but had evidently made no impression.[40]

Two days later, on October 21, the Arab League's Bureau for the boycott of Israel called on its members to implement punitive measures, to boycott Dutch firms and to cut off the oil lines to Rotterdam. This request, an official of the Bureau informed the world, was in response to the wholly pro-Israeli line taken by The Hague.[41] October 21 was a black day for The Netherlands. Iraq nationalized the Shell share of an Iraqi oil company, and Syria called for a general boycott of The Netherlands and specifically of KLM. The same day, Jordan declared itself ready to boycott KLM.

But that was not all. Algeria implemented the boycott of the Arab

League's Boycott Bureau the very same day. In a conversation with the head of the Western Europe and North American Bureau, Ambassador G.W. Bentinck in Algiers was given *eight reasons* why it been decided to impose a boycott. The first reason was the government's pro-Israeli declaration immediately after the outbreak of the war. The second was the pro-Israeli stance taken by Van der Stoel in his meeting with the four Arab Ambassadors. Thirdly, The Netherlands had adopted a pro-Israeli posture in the UN and in other international bodies. Fourthly, the government had supported Israel within the EC; and furthermore, the government had approved the recruiting of volunteers and had offered the transport facilities of KLM.

It also struck Bentinck that his opposite partner in this conversation was remarkably well-informed of the deliberations of the *Comité Politique* that preceded the EPC meeting in Copenhagen. One of the EC's member states, probably France or Italy, had evidently allowed the content of these discussions to leak. The last three Algerian arguments were not directly connected to the war, but served rather to demonstrate that The Hague had generally adopted a pro-Israeli or pro-Jewish stance. For example, the Dutch Foreign Ministry had summoned the Austrian Ambassador to protest against the announced closure of a refugee camp for Jewish emigrants from the Soviet Union. And finally, Den Uyl had openly declared that he suffered sleepless nights thinking about the Jewish victims of the Second World War. The embargo, Bentinck was told, would be swiftly followed by other Arab countries, but otherwise it was hoped that Dutch exports to Algeria would not suffer as a consequence. A few days later these arguments were reiterated in a letter from the Algerian President Boumedienne to Den Uyl.[42]

The Algerian move was a serious threat because it could possibly be followed by similar measures from other Arab OPEC countries. On October 21 Kruimink, the Co-ordinator of Intelligence and Security Services, informed Den Uyl and Van der Stoel that following their audience with Van der Stoel, the Arab Ambassadors in The Hague, as we have already seen, had advised their governments to impose an embargo if the Dutch Cabinet continued its anti-Arab policy and its military support for Israel. It would seem that the Arab diplomatic codes had been cracked.[43] A day later, Kruimink further warned Van der Stoel that it was learned 'from a reliable source' that the Algerians were actively prevailing on other Arab countries to follow their example. It appeared that Kuwait had already decided to follow suit, thought Kruimink, and the next day confirmed his impression.[44]

A day after the Algerian decision, the Dutch Ambassador reported from Damascus that Syria had also decided to follow Algeria. A junior minister had informed him that there had been a meeting of Arab Ambassadors in Brussels at which it had been decided to advise their respective governments to declare a total boycott of Dutch interests. The embargo should be put into operation, since it was apparent that The Hague had not modified its attitude, deemed so prejudicial to the Arab world.[45]

Nationalization in Iraq

In spite of this development, the Arab OPEC countries proved incapable of any unity of resolve. For different reasons, Iraq and Iran did not join the embargo.[46] At the OAPEC conference in Kuwait, Iraq had adopted a hard line: complete liquidation of American assets in the Arab world, withdrawal of all Arab deposits from the USA and a general hardening of the political attitude toward Washington.[47] Subsequently, Saddam Hussein would explain that although the embargo against the USA and The Netherlands was just, it was far too weak a measure. Restricting production only succeeded in damaging Arab business. The oil weapon had to be aimed effectively against Washington and crucial American interests.

Iraq announced on October 21 that the Dutch share – i.e. the Shell share – in the Iraqi *Basrah Petroleum Company* was to be confiscated and nationalized, or rather the Dutch part (60%) of Shell's participation. Two weeks earlier, on October 7, immediately after the outbreak of the war, Iraq had nationalized the interests of the American companies Exxon and Mobil Oil in *Basrah Petroleum Company* as a first step in an 'oil battle against Israel and its supporters'.[48]

Explaining the decision taken against Shell, reference was made to the 'aggressive attitude toward the Arab nation' and 'support for our enemy'. A statement from the Iraqi press agency laid a whole series of accusations at The Hague's door, the grievances already known. In particular, Dutch territory was being used as a bridgehead for assistance to Israel.[49] Although the Iraqi grievances were essentially political, Shell initially set no great store by any Dutch governmental move, preferring to take independent action against the nationalization decision themselves. Nevertheless, it became clear to one of Shell's top men several days later that it would be exceptionally difficult to get the nationalization revoked. His Iraqi counterparts at negotiations were extremely negative over the hostile Dutch position, leaving the Shell representative to conclude that perhaps diplomatic steps might procure compensation for the nationalization.[50]

Nonetheless, for several days there was uncertainty in The Hague over the Iraqi moves. On October 22 a spokesman let it be known that news of the nationalization had only reached The Hague via the news media. In spite of the lack of information, the Dutch government regretted the measure: 'The Netherlands observes friendly relations with Iraq and moreover values their continuation.' The Cabinet hoped that all misunderstandings could be cleared up, a vain hope as soon became apparent: the situation in Iraq remained alarming. It was reported from Baghdad that a boycott of KLM was in the wind.[51]

When a Shell delegation held new discussions in Baghdad the following week, it was once more concluded that the nationalization had been a political decision. This was evident not least from the fact that the Oil Minister Saadun Hamadi was not fully informed of what was happening. During these discussions, a hard line was taken by the Iraqi side, who again referred to the hostile attitude of The Hague. As one Shell director confided, it appeared highly improbable that the decision would be revoked in the foreseeable future. On the question of compensation, too, the Iraqi attitude was far from compliant.

Van der Stoel, on the other hand, was of the opinion that the political justification for the nationalization 'should be taken with a pinch of salt'.[52] This was an understandable view, since the nationalization in Iraq was part and parcel of a whole series of comparable measures, both in Iraq and elsewhere in the Arab world. Van der Stoel thought the Dutch attitude to the October War was more likely a convenient stick to beat them with, and that this was also the case with the embargo. Shell, however, maintained the view that in Iraq it was a matter of political action, a reaction to The Hague's Middle East policy. A member of the Shell delegation that had negotiated to no avail in Baghdad was pessimistic on his return to London, telling Ambassador Gevers that the embargo looked like being a long business. It was important to keep talking to the Arabs, but he saw the situation worsening rather than taking a turn for the better.[53]

A New Government Statement

It was now a matter of priority to try to prevent the embargo spreading further and to head off other anti-Dutch measures. The Saudi position was most important because around one-third of Dutch oil imports came from Saudi Arabia. On October 22, Director-General of Political Affairs of the Dutch Foreign Ministry, Van Lynden, had a meeting with the Saudi Ambassador, Rashad Nowilaty, at which hope was expressed on the Dutch side that King Feisal would not join an embargo. Van Lynden sought to clear up the many 'misunderstandings' between the two parties and in fact was partly successful. Nowilaty promised that his country would continue to supply would-be customers as long as this incurred no principles that might damage Saudi Arabia. He promised to inform King Feisal of the various currents at play in The Netherlands and also to relay expressions of sympathy for the Arab cause. Van Lynden afterwards concluded that it had been a good meeting. He had the impression that the Saudis were in no hurry to join the embargo, though the pressure exerted on them to do so should not be underestimated.[54]

With regard to the Algerian embargo, it was decided by Van der Stoel to make no formal protest but to react, as it was termed, 'with dignity and moderation'. Ambassador Bentinck was instructed to respond with a demarche, to the effect that The Hague regretted the Algerian decision, all the more that it was based on incorrect information. The Jewish people and Israel perhaps enjoyed much sympathy in our country, but the government's policy with regard to the Middle East was 'balanced' as was evident from its stated view that a cease-fire must be acceptable 'to both parties'.[55] These instructions were also sent to various diplomatic posts in the other Arab countries that they might approach the governments on the spot in the appropriate way. The Algerian temporary *chargé d'affaires* was summoned to Foreign Affairs to draw his attention once again to The Hague's point of view.

For all these efforts, however, The Netherlands was about to find itself in serious difficulties. The situation called for close contact with the major oil companies. On October 22, Van der Spoel spoke with the chief executive of the Royal Dutch Shell Group, G.A. Wagner, who had requested the meeting mainly as a result of the nationalization of the Shell interests in Iraq.[56] Although F. Grünfeld, a Dutch expert on the Middle East, claims that Shell did not try to exert any influence on foreign policy[57], according to the Dutch Foreign Ministry report of that meeting, Wagner did make substantial allusions to government policy. Wagner emphasized that Shell kept itself out of Middle Eastern politics, even though the sympathies of the board lay with Israel. It seemed undesirable to Shell's president that his country should be too much identified with the USA by the Arab countries. Because Shell particularly feared that the nationalization in Iraq would be emulated by other Gulf States such as Abu Dhabi, Qatar and Oman, Wagner said he would be grateful to the Cabinet if, wherever possible, the emphasis could be laid on the 'even-handedness' of its Middle Eastern policy.

Van der Stoel, according to his own testimony, was doing everything he could to refute the rumours of support given to Israel, but so far this effort had met with little success. He promised to summon the Iraqi Ambassador. Wagner appeared not entirely reassured and suggested making a statement expressing commiseration with the fate of the Palestinian refugees in order to strike a more suitable chord with the Arab world. Van der Stoel, however, did not comply with this suggestion. Finally, Wagner expressed his concerns over the consequences of the price rises, particularly for the developing countries, arguing that consultations must be initiated as a matter of some urgency. According to the Minister, this should be handled within the framework of the OECD.[58]

Meanwhile, a new governmental statement was being prepared, duly released on October 23, which spoke of the 'misunderstandings' that had arisen concerning the Dutch position with regard to the Middle East conflict. A solution would have to include all elements of resolution 242. The statement cited the recently presented explanatory memorandum on the Dutch Foreign Ministry budget, in which it was asserted that Israel would have to withdraw behind 'secure boundaries to be agreed, to a territory approximately of the same area as her territory before the Six Day War.' In addition to which a peace settlement would have to be based on the sovereignty and territorial integrity of all states in the region. Furthermore, a solution would have to be found for the refugees.[59] This contained little that was new. Nonetheless, the Shell board expressed their thanks for the new government statement, while Shell's Arabists began preparing a translation.[60] This statement, which was sent out on October 26 by the Dutch Foreign Ministry to Ambassadors in the Arab countries, was an example of effective collaboration between the Ministry and Shell.

The Embargo Spreads

Despite all diplomatic activities, the embargo spread, with Kuwait following the Algerian example on October 23. In explanation, the Dutch temporary *chargé d'affaires* in Kuwait, D.M. Schorer, was referred to his government's pro-Israeli attitude, and in particular to the presence of Vredeling at a pro-Israeli demonstration. Schorer reported that Oil Minister Abdel Rahman al-Atiki had said that the damage could only be repaired if The Hague broke off diplomatic relations with Israel. This, according to Schorer, was also representative of Kuwaiti public opinion.[61] Following Kuwait there came embargo announcements from Abu Dhabi and Qatar, and finally from Oman, while from another quarter came rumours that Libya was also about to join the embargo; and indeed, on October 30 this happened.

At this stage, the Dutch Foreign Ministry was under the impression that Algeria was playing a major role in spreading the embargo. Attempts to mollify Algiers came to nothing. On October 24 Bentinck offered the new government statement to the Algerian Director-General for Political Affairs, who promptly blamed The Netherlands for their

> evidently constant support for Israel .... without ever compensating for this attitude with the least appearance of understanding or support for the Arab or Palestinian cause.

The Director-General later declared that relations could only be repaired if The Hague adopted the same point of view as its EC partners.[62] The Algerian Ambassador, Messaoud Aít Chalaal, later explained that The Hague's attitude in NATO had aroused the Algerians' ire. Of all the NATO countries, only The Netherlands had unreservedly agreed to allow overflights of American transport planes to Israel.[63]

On October 31 a letter signed by Den Uyl was delivered to the Algerian President Boumedienne – his reaction to the missive which had announced the embargo. Den Uyl wrote that he shared the Algerians' concern over the conflict in the Middle East. On the outbreak of the war, the Dutch Premier emphasized, The Hague had urged an immediate end to hostilities. He went on to underline the fact that The Netherlands had always pressed for a peaceful resolution of the conflicts in the Middle East on the basis of resolution 242.[64] But this letter too was for the time being of no help.

Although the embargo was spreading, there had been from the outset no united Arab front. This was evident, for instance, from the fact that various Arab countries had only joined after several days. There were other Arab countries which did not join at all, such as Iraq. Iran also remained aloof, though from a much more moderate standpoint than Iraq. The Dutch Ambassador in Tehran reported that the Shah was trying to take a softer line, declaring that it was unfair of Saudi Arabia to punish Europe for the American position.[65]

In Cairo, Dutch Ambassador F. Von Oven reported that the Egyptian government was also unwilling to take part in the embargo, certainly as long as there was discord between the Arab nations. Egypt was a modest oil exporter, which might account for the Egyptians' reluctance. In any case, Cairo considered the embargo an overly hasty action.[66] Later, Egypt's temporary *chargé d'affaires* in The Hague, Mohammed Said El Sayed, explained that Van der Stoel's meeting with the four Arab Ambassadors had not been the cause or the main reason for imposing the embargo. He had found the exchange reasonable and constructive.[67] This judgement would have confirmed Van der Stoel in his conviction that the embargo was not essentially a reaction to his policy, but had in all probability been in preparation for some time.

As the biggest oil producer, Saudi Arabia was of course more important than Egypt. At this stage, Saudi Arabia also declined to join the embargo, though very soon reports came of steps in this direction. On October 24 the Dutch First Secretary at the Embassy in Jeddah reported that an embargo decision had 'already been prepared'. It was still possible, he learned, to turn the tide in Saudi Arabia through some positive deed – for instance if The Hague were to declare that it believed 'that resolution 242 should now be implemented without further delay'.[68]

Two days later the possibility for such a 'positive deed' seemed to present itself when Saudi Arabia gave The Hague an ultimatum, delivered by Ambassador Nowilaty on October 26 to the Dutch Director-General for Political Affairs, Van Lynden. The ultimatum contained three demands that must be satisfied if implementation of the embargo were to be averted. Firstly, the Dutch Cabinet must condemn the Israeli aggression. Secondly, it must demand Israel's withdrawal from the occupied territories[69] and thirdly, the Palestinians' right to self-determination must be recognized.[70]

The Dutch Council of Ministers met the same day to discuss the ultimatum. Meanwhile, it was learned that Saudi Arabia was classifying consumer countries according to three categories: friendly, neutral and hostile states, with The Netherlands in danger of finding itself ranked as a hostile state. Van der Stoel concluded that there was no point in yielding to the ultimatum. In his view, The Hague was being used as a scapegoat, mainly, he suspected, to strike at The Netherlands' transit and refining capacity and, as result, to put pressure on the whole of Europe, although of course this was not to deny that there was considerable sympathy for Israel among the Dutch population.

During this ministerial council, though in a circumspect manner, a difference of opinion became public between Lubbers and Van der Stoel. Lubbers expressed his great concern over the situation that had arisen. There was now a danger that the oil stream might shift from Rotterdam to other countries. He believed that some Arab countries saw Washington and The Hague as too much on the same wavelength. He thought it was probably France that had depicted the Dutch position in the EC as bluntly pro-Israeli, but he was also of the view that the Cabinet itself was partly responsible for this image. He advocated that the main determining principle in the current situation should 'not be the aggression of the Arab states' but rather 'the resolution of the conflict on the basis of resolution 242'.

Van der Stoel replied that his policy was aimed at removing misunderstandings. During the General Debate in the Second Chamber, support for Israel may have been adopted as a clear principle, but that had been with an eye to public opinion. Dutch policy continued to be based, as before, on the implementation of resolution 242 as the condition for any Middle East solution. According to Van der Stoel the government must demonstrate publicly that there was no panic. He was not pessimistic. An attempt to provide a counterweight to the growing criticism of the radical Arab countries had to be essayed, particularly through Saudi Arabia and Egypt.

Lubbers replied that, given the way the Saudi position had been formulated, there was apparently room for talking, and he warned against reacting too negatively to the ultimatum. Van der Stoel agreed, but at the same time felt that it had to be made clear that the Dutch Cabinet could not be blackmailed. In any case, in conversation Ambassador Nowilaty could be referred to the declaration of October 23, which stated that Israel must withdraw to roughly the same territory as before the 1967 war.[71] In the end the Cabinet decided not to comply with the ultimatum. Van der Stoel would ask Nowilaty to call on him, declare his failure to understand the content of the ultimatum, and would point out that the government also failed to understand why The Netherlands was being treated differently from other EC countries. On October 27 Nowilaty was summoned to an audience with Van der Stoel at which he was handed a brief statement referring once again to the government statement of October 23. In the Dutch view, there were a number of misunderstandings, and therefore the Cabinet sought further diplomatic consultations.[72]

There followed several confused days during which it quickly became apparent that this response was inadequate. According to some reports, the Dutch reaction immediately provoked Saudi Arabia to join the embargo, though other reports seemed to point the other way. The Dutch Ambassador in Jeddah, in a telephone conversation with a journalist from the *Algemeen Dagblad* on October 31, stated that in his view Saudi Arabia had not called an embargo, a judgement which the paper published the same day.[73] The Dutch Foreign Ministry was not at all happy with this statement and promptly instructed the Ambassador to make no further public pronouncements over the embargo, since these could very well prompt the Saudi government to actually join the embargo demanded by the Arab League.[74]

In The Hague, meanwhile, efforts were made to maintain friendly relations with the so far apparently moderate countries, Tunisia and Iran. To this end, Van Lynden met with the Ambassadors concerned, trying yet again to remove the 'misunderstandings' that had arisen. He further put it to the Iranian Ambassador that the embargo was in conflict with the conditions of GATT, and observed that his government in its reaction had for the present taken a dignified and moderate attitude. Given the position of Rotterdam, imposing the embargo meant striking at the whole EC. Van Lynden emphasized his appreciation of the position taken by Iran, which was continuing to supply oil and was not using it as a political weapon.

Van Lynden expressed himself similarly on October 30 to the Tunisian Ambassador, who let it be known that his country indeed took a moderate position, but out of solidarity had to conform with the other Arab states. At that moment, Tunisia was not able to oppose the voice of the Algerians. In his view, the Arabs had been most incensed by the Dutch government's statement of October 9, which spoke of withdrawing behind the truce lines existing before the October War. These were, after all, not recognized state boundaries but demarcation lines resulting from the illegal, six-year Israeli occupation of Arab territory.[75]

On November 2, the move that had been feared was finally taken: Saudi Arabia definitively joined the embargo. There had been uncertainty over Saudi intentions for several days. Kissinger later maintained that Saudi Arabia had not really themselves taken a decision over the embargo but 'may have felt their hand tipped by published reports that the SAG (the Saudi Arabian government) had or was about to take such a decision'.[76] This view corresponds with Van der Stoel's own understanding that it was public speculation over the institution of a Saudi embargo that had contributed to its actual implementation.

In reaction to these developments, Van der Stoel instructed diplomatic representatives to refrain from asking for further clarification in future, either over the oil embargo or any other measures taken against their country. The attitude of The Hague was to be one of 'dignity with moderation'. The impression must at all costs be avoided that there was any confusion or panic. Low key, low profile were to be the watchwords for the way The Hague would conduct itself, while all would be done to ensure business as usual. The developments in Saudi Arabia demonstrated, according to Van der Stoel, that public pronouncements over the embargo of Dutch and American business had done no one any good.[77] The Dutch government and Foreign Ministry did not adopt a completely passive position. Nowilaty in particular was at the same time requested, in secret, to see whether a special legation would be received by King Feisal. This Van Roijen mission would in any case have to wait a while.[78]

The De Lavalette Mission

On October 26, in order to save what could be saved, the Council of Ministers agreed with the proposal of the Ambassador in Tehran, P. Renardel de Lavalette, to arrange a tour of the Gulf States, i.e. Iraq, Kuwait, Qatar and the United Arab Emirates. The preparations for such a tour had already been underway for several days in close consultation with Shell, who provided the Dutch Foreign Ministry with extensive information concerning Shell's interests in the relevant countries. It was also Shell which suggested the sequence order of visits, although this was not entirely adhered to. During his tour, De Lavalette always had close contact with the Shell representatives on the spot.[79]

Although the Dutch Cabinet was later to complain of a lack of solidarity between the EC countries, this initiative was set up without the knowledge of other EC partners. Only London was informed in advance, for the reason that the English oil interests were closely bound up with the Dutch. Within the EC, however, no other member state was told by Van der Stoel what was planned. A modest press conference would be held on the day of De Lavalette's arrival. Accordingly, the plan had to be communicated in strictest confidence to the British government.[80]

De Lavalette would have to travel to the Gulf States to cultivate goodwill. His most important task was to remove misunderstandings and false impressions. In connection with the anger allegedly caused by Vredeling's presence at the pro-Israel demonstration in the Amsterdam Bourse, Van der Stoel instructed him to emphasize that although Vredeling may have been there, it was purely in a private capacity. De Lavalette should anyway 'not return to this point of his own accord'.[81]

It was soon clear, however, that De Lavalette's mission was not a simple one. In preparing his trip, he spoke in Jeddah with the Ambassadors of the countries he intended to visit, who informed him that their governments were particularly furious over Vredeling's attendance at a 'Zionist demonstration'. They did not accept the argument that this had occurred 'in a purely private capacity'.[82] Preparing for De Lavalette's mission in Kuwait, Ambassador Schorer was told the same: The Hague's viewpoints, Vredeling's presence in the Bourse and clandestine support for Israel did not contribute to good relations. Besides, it was said, the Arab world was not asking for an anti-Israeli position, but it did demand an objective approach.[83]

The journey began badly. When De Lavalette arrived in Kuwait on October 27, he did not manage to arrange an interview with a single relevant authority. In desperation, the Ambassador journeyed on to Abu Dhabi where he had a meeting with the Under-Secretary for Foreign Affairs. This went rather better. The Under-Secretary informed him that the Emirates had hesitated several days over the implementation of the oil embargo, but had in the event been unable to escape joining the Arab line. After Abu Dhabi came Qatar. The meeting there with the Foreign Minister, M. Suheim, began in unpleasantness, but this was later checked, with the Minister going so far as to express regret over the embargo and promising that he would inform the Emir of his conversation with De Lavalette.[84]

Back in Kuwait, De Lavalette was granted a meeting with the Under-Secretary for Foreign Affairs, Rashid al-Rashid. Following several emotional accusations, which De Lavalette suspected were intended for their effect in the subsequent report, Rashid admitted that the embargo had been introduced on the basis of the 'atmospherics' rather than any solid evidence that The Hague had actually breached the principle of neutrality. In Kuwait too, the sore point was raised of Vredeling's action. As an illustration of Kuwait's attitude, Rashid pointed to the great number of Palestinians working in Kuwait. He then took up Van der Stoel's point that the embargo was actually aimed not so much at The Netherlands as the whole of Europe. Rashid found this argument unconvincing. The nine EC member states, after all, wanted oil as normal and therefore desired no part in the Dutch-Arab conflict. The Hague therefore stood alone.[85]

Subsequently, De Lavalette held conversations with the Foreign Ministers of Oman and Bahrein on November 2 and 3. Both Arab Ministers gave the impression that their countries had been more or less forced to join the embargo. In Oman, moreover, De Lavalette was reassured that in all probability The Netherlands would manage to survive through the 'rescheduling' of oil.[86]

The last country on De Lavalette's round was Iraq, where he was received in an 'unexpectedly courteous' fashion, yet at the same time was given to understand that the nationalization of oil assets was a legal fact and could no longer be reversed. De Lavalette was given the impression that Iraq was not immediately about to implement an embargo.[87]

In his final report, De Lavelette judged that the embargo had not been implemented with full conviction by several Arab countries, but that these had been more or less corralled out of solidarity. The basic cause, he had learned from his conversations, lay in the inadequate attention given by The Hague to the Arab side of the Middle Eastern conflict. This one-sided approach was coupled with extreme pro-Israeli pronouncements – highly distressing to the Arab countries – of certain Dutch authorities. De Lavelette had the impression, therefore, that friendly words addressed to the Arab countries could do much good.

The question, thought De Lavalette, was how effective the embargo was. Kuwait was apparently the sole country that had detailed plans for its effective implementation. De Lavalette was in general not pessimistic. He considered it unlikely that there would be further punitive measures taken against The Netherlands. His general impression was that matters were sure to simmer down, although the initial Egyptian military success had created a huge stir in the Arab world, and the oil weapon was now seen as a powerful resource. De Lavalette concluded that the countries he had visited would follow the Arab line, but indicated 'that although the embargo might be formally adhered to, what further happened to the oil was otherwise a matter of no concern'.

This was an important and reassuring conclusion. 'The Netherlands would manage to survive', was the assumption in the Arab countries, specifically through 'rescheduling' the oil stream to Rotterdam. If this were indeed the case, De Lavalette stated, it would be best for various reasons 'if it were revealed as little as possible to the outside world'. If it were known, 'the shadow play would lose its value for the Arabs, and that could only serve to provoke new actions'.[88]

It was also assumed in the USA that the embargo was far from effective. The American Ambassador in Tripoli was already able to report on October 25, 1973, that the Libyan authorities had let it be known that they had no objection to exporting oil 'not documented as destined directly for US ports'. It turned out that the Libyan National Oil Company was simply continuing to fill American tankers or oil tankers bound for the USA without showing the least concern for the question of the oil's destination. The National Oil Company turned away not a single American buyer, so long as they were prepared to pay high prices.[89]

It became apparent from other sources that even Algeria, the main instigator of the embargo, was not taking it too literally. On October 30, the American Ambassador in Algiers was able to report that the Algerian government evinced no serious interest in the ultimate destination of their oil exports. A captain of a tanker originally bound for the USA, but who had in the meantime altered his destination on the freight documents, was informed by the Algerian official that his government had absolutely no interest in what tankers did once they were on the open seas. In addition, there still remained the question of whether Algeria would reduce its oil production by 10% as agreed within the OAPEC.[90]

None of this, however, could disguise the fact that, however merely formal, an embargo had been declared against The Netherlands. It was abundantly clear that there was great irritation in various Arab countries with the attitude taken by The Hague. Ex-minister Luns, at that time Secretary-General of NATO, had also noted this whilst in Turkey between October 24 and November 1. According to Luns, this resentment was not so much a consequence of what The Netherlands had in fact done to favour Israel, but the public remarks of some Cabinet members. Luns pointed out to his interlocutors that The Hague had accommodated the Arab countries, in spite of the pressure brought to bear from 'certain quarters', by declining to supply Israel with any war materials. The present government had difficulties, according to Luns, because of the role of left-wing 'agitators'. He assured them that Van der Stoel was certainly not anti-Arab.[91]

The Second Chamber

Meanwhile, the seriousness of the embargo began to penetrate through to public opinion and the Second Chamber, leading in turn to a remarkable change of stance. Immediately after the outbreak of the war, the Second Chamber stood foursquare behind Israel. This now began to change, and a major role in this change was played by the party-political composition of the Den Uyl Cabinet, so recently put together. It was suggested by the conservative-liberal VVD that the government and Van der Stoel with their politics of 'bearing witness' had put Dutch interests in the Arab world at risk. Criticism of government policy was also expressed in the press, particularly NRC *Handelsblad* and *De Telegraaf*. NRC *Handelsblad* even spoke in somewhat overwrought terms of the 'collapse of our step by step, almost completely democratized, foreign policy'.[92] Rejecting this criticism at a press conference held on October 30, Van der Stoel insisted that he had never taken an anti-Arab standpoint.

The standpoints taken on the October War and the Arab grievances were set out once again in an internal Foreign Ministry memorandum on the eve of a debate in the Second Chamber. The Dutch government's position was in general balanced; Israel had to withdraw from the occupied areas, subject to minor border corrections; Israel's security had to be guaranteed; a just solution had to be found for the position of the Palestinians. Beside this, after the outbreak of the October War, The Netherlands had never spoken of aggression against Israel. The EC declaration of October 13 was unanimously accepted and embraced all the elements of resolution 242. The only objection the Cabinet had was to a mandate granted to Great Britain and France to speak on behalf of the Nine in the Security Council.

The Dutch Foreign Ministry memorandum denied most emphatically that The Hague had given Israel military support. There had been no recruitment of volunteers, and The Netherlands had not served as a bridgehead for arms transports to Israel. Since the outbreak of the war, no permits had been granted for the export of arms to the war area. KLM had transported neither soldiers nor material to Israel. Only on October 9 had there been a further flight to Tel Aviv. No oil had been supplied to Israel, and Vredeling had participated in a pro-Israeli demonstration 'in an entirely personal capacity'. It was true that the Dutch people, for historical reasons, had much sympathy for the Jewish people and for Israel, but during the war there had also appeared expressions of understanding for the Arab cause in the national press.[93]

On October 31, the Dutch government's policy with regard to the October War was once more raised in the Second Chamber for further extensive discussion. The VVD spokesman, F. Portheine, opened this discussion by putting several critical questions. He thought perhaps the reporting had been rather haphazard and asked whether the De Lavalette mission that was underway could not be used to lay the ground for a visit by Van der Stoel to the Arab countries. M.W.J.M. Peijnenburg (KVP) also thought that Dutch diplomacy had not always been effective given the number of misunderstandings that had apparently arisen between The Netherlands and the Arab countries. More fundamental was the criticism of M.J.J.A. Imkamp (of the D'66 party) who asked whether the government was finally ready to underwrite the French interpretation of resolution 242. The Cabinet was mainly supported by A. van der Hek (of the Labour Party, the PvdA). In general, the Chamber seemed to have a better understanding of the Arab side and the position of the Palestinians than they had three weeks earlier.

In their replies, Den Uyl and Van der Stoel emphasized that government policy had not altered since the statement of October 9, on which occasion the Chamber had scarcely uttered a word of criticism. Van der Stoel went into the background of the embargo. The Arab countries felt that The Netherlands had manifestly not understood them for a long time, thought the Minister. On the Dutch side, despite sympathy for Israel, regular initiatives had been essayed to try to remove this feeling. In order to improve relations with the Arab countries, Van der Stoel had invited the Egyptian Foreign Minister to visit The Hague at the end of August. Dutch diplomats had used every possible opportunity to overturn the many misunderstandings current in the Arab capitals. Van der Stoel emphatically declared his conviction that the meeting with the four Arab Ambassadors had not been the cause of the embargo. The decision had in all probability been taken earlier, and it was directed against the whole of Europe. He endorsed the standpoint that the Palestinian question also deserved attention.[94]

The Cabinet was thus subjected to considerable criticism during this debate. During the uncertain days immediately preceding the Saudi decision to join the embargo, this disapproval was also voiced outside the Chamber. The PvdA member of the Dutch parliament, R. ter Beek, publicly stated on November 2 that, in view of the often divergent and conflicting reports concerning the attitudes and aims of various Arab countries, particularly Saudi Arabia, he had serious doubts over the quality of the diplomatic service. At a press conference on the same day, Den Uyl rejected this criticism: Dutch diplomatic representatives in Arab countries, he said, had nothing to reproach themselves with.

KLM

The measures taken against The Netherlands were not restricted to oil. From the moment the October war broke out, rumours began to circulate to the effect that the USA and Israel had arranged supplies of arms by means of hired European aircraft. As we saw, KLM in particular was supposed to have transported military experts and volunteers to Israel. These and similar accusations were put out by – among other sources – the Arab League's Bureau for the Boycott of Israel. It quickly became clear that various Arab countries were considering denying KLM landing rights.

KLM denied all involvement from the outset. In a press statement it was firmly stated that since the outbreak of the October War, there had been no more flights to Cairo, Tel Aviv, Damascus or Amman. There had been one more flight to Tel Aviv on October 9 to pick up stranded passengers, but that was all. The Dutch Foreign Ministry sought to back up KLM. Ambassadors in Beirut, Damascus, Cairo and Jerusalem received instructions to communicate the contents of the KLM press conference to the respective authorities without making any public statement.[95] In spite of this, KLM's position became more difficult when accusations began to emerge over involvement in arms transport to Israel via Schiphol (Amsterdam Airport).

It has also not been excluded that KLM might have hired aircraft to the Israeli airforce or to El Al. An ex-member of the Mossad, the Israeli intelligence service, believes that KLM flew mainly from Tehran to Tel Aviv. At the time, like Israel, Iran used weapons systems almost exclusively of American manufacture. The Shah would have sent war materials to Israel through Turkish air space, for which purpose KLM transport planes would also have been used. Former officers of the Dutch secret security service claim that the Dutch Schreiner Airways and other charter companies were involved in the secret arms deliveries via Schiphol to Israel.[96]

On October 16 the Dutch Ambassador in Damascus, J.B.E.Ph. van Hoeve, reported that the Syrian government appeared to be endorsing one of the accusations against The Netherlands. According to Under-Secretary for Foreign Affairs Rafai, 'fifteen aircraft loaded with weapons and ammunition left Schiphol'.[97] On October 21 the Syrian government called for a total boycott of The Netherlands, including KLM. Van der Stoel maintained his policy of not reacting publicly to the accusations levelled at KLM, but trying rather to resolve the matter as far as possible through diplomatic contacts. Diplomatic efforts to support KLM, however, made difficult headway. In Cairo, Ambassador Von Oven was summoned by the acting Foreign Minister, Ismail Fahmi, whose communication that Dutch volunteers taken prisoner by Egypt would not be considered as prisoners of war seemed to indicate that he gave little credence to Van der Stoel's disclaimers. And yet Von Oven came away with the impression that Fahmi wanted to give the Dutch standpoint on the Middle East the benefit of the doubt. The Boycott Bureau and the Middle East were not totally convinced by the KLM statement, thought Fahmi. He offered to get both statements broadcast again on radio, TV and in the press, an offer that was gratefully accepted by Von Oven.[98]

It became clear on October 21 that Jordan was willing to join the KLM boycott demanded by Syria. The Jordanian Director-General for Political Affairs informed the Dutch Ambassador, A.C. Vroon, and a KLM representative that in view of the communal struggle against Israel his country, to his regret, was compelled to follow Syria. Action in Amman was pointless: The Netherlands would have to present evidence to the contrary to the Boycott Bureau.[99] But apart from this, Van der Stoel had forbidden any contact with this Bureau, since it was not recognized by the Dutch government. Two days later, news came that Iran was also about to join the boycott of KLM.

KLM did not let matters rest there. On October 23 a letter was composed in Arabic on behalf of the President-Director in which issue was taken with 'untruths'. KLM succeeded in making direct contact with the Boycott Bureau and in seeing that no new steps against KLM would be urged. A request to withdraw all punitive measures, however, would have to come from Damascus where, meanwhile, KLM had already approached the Syrian Foreign Minister. [100] On October 25 the Dutch Ambassadors to the Arab countries once more received instructions to stand behind KLM as far as possible.

On October 26, however, the Arab League's Boycott Bureau in Damascus let it be known that KLM, together with Sabena, Lufthansa and Air France, had been informed they could only resume flights to Tel Aviv after Israel had vacated the occupied territories. This seemed a good occasion for cooperation, but KLM was not in favour of joint action, as it informed Van der Stoel. KLM's position in the Middle East was stronger than that of Sabena, Air France or Lufthansa,[101] in spite of all the problems, and for this reason the Minister refrained from any initiatives involving joint action.

In fact, aside from the Syrian appeal, it turned out that only the Jordanian government had announced a boycott against KLM. In Amman, Ambassador Vroon tried to get this action undone at the end of October, with the Jordanians promising that KLM would be allowed to resume flights to Amman once it became clear that Damascus gave permission to fly over Syrian territory. The Ambassador in Syria, in accord with the 'low key policy', made no direct approach but merely assumed that KLM would be able to land again as soon as the Damascus airport was opened to civil traffic.[102]

The situation was thus a confusing one, but apart from Jordan there was no question of actual boycott. This did not altogether prevent the situation starting to become more difficult for KLM at the end of October, when boycotts were called by unions and other organizations for which the governments of the countries concerned could not really be held responsible. In Iraq, such actions were staged by the Association of Iraqi Travel Agencies. In Libya, a workers' boycott of American and Dutch ships and air traffic was announced by the Federation of Arab Trade Unions. On November 2 the International Federation of Arab Trade Unions adopted a severe resolution which, among its articles, recommended a total boycott of American and Dutch ships, aircraft and goods; and in fact in Libya there were boycott actions against KLM. In Egypt, where the political situation was tense following the collapse of the Sinai front, similar actions were threatened.[103]

Conclusion

Low profile or not, the situation looked particularly unpropitious for The Netherlands, although the Dutch were not entirely alone. Denmark was also subjected to an embargo. But that was of small comfort. Those states that had proclaimed an embargo against The Netherlands provided more than two-thirds of total Dutch oil imports. If these were really going to be blocked, this would undoubtedly entail complex economic consequences. Diplomatic measures taken so far had had relatively little effect. Van der Stoel was still assuming on October 6 that it would be possible to put up some resistance to Saudi Arabia and Egypt, but a week later the biggest oil producer of all had joined the embargo. In the meantime, the Minister had made it clear to his representatives that the policy must be low key, hoping to conduct business as usual as much as possible.

It is not easy to establish what was the actual purpose of the Arab countries in choosing The Netherlands as their target. The reason that was served up by all the Arab states involved for imposing the oil embargo was the alleged support for Israel during the October War, but the question remains whether the embargo was really – or solely – about Dutch Middle Eastern policy. By striking at Rotterdam – and Van der Stoel was right in this – the whole of Northwest Europe was affected because of the port's transit importance in Europe's oil flow. The embargo therefore looked like an attempt to put pressure on the whole EC. On the other hand, the Arab measures against The Netherlands rapidly led to serious conflict within the EC, which only seemed to make any EC political involvement in the Middle East all the more unlikely.

But there was certainly more at stake during the oil crisis than the Israeli question. The measures taken by the Arab OPEC countries were not a purely political sanction against those countries that had allied themselves with Israel during the war. The oil crisis was also part of a struggle in the international oil sector. The Arab countries, albeit from different perspectives and interests, had set themselves to break the mould of traditional relations within the oil sector, in which Rotterdam occupied an important position. It was the home port of one of the *Seven Sisters* and furthermore a crucial link in the chain of processing and distributing oil and oil derivatives in Northwest Europe. So at a quite early stage of the crisis, Van der Stoel was convinced that the embargo had already been prepared months before the outbreak of the October War.

In conversation with Golda Meir on November 13, Den Uyl also emphasized that the oil weapon had been planned well in advance. PvdA leader in parliament Ed van Thijn, learned of this on November 15 in the Permanent Parliamentary Committee for Intelligence and Security Services. The parliamentary leaders of the four main political parties were told that as early as July there had been indications that an embargo had been decided on. According to Van Thijn the Dutch attitude had little to do with it. The same statement was repeated the following day in the governmental consultation between the PvdA Ministers and parliamentary party leaders.[104] These judgements do not square at all with the assertions of the ex-honorary consul for Kuwait, Mahmoud Rabbani, to the effect that the decision to boycott The Netherlands was of Van der Stoel's fault.[105]

However, when it came to the boycott and other measures, such as nationalizations in Iraq, there were differences of outlook, objectives and expectation within the Arab world. [106] For some, the Palestinians for example, the primary objective was political, part of the fight against Israel. For others, the struggle was to break out of existing power and property relations in the international oil sector, for example Iraq and possibly also Algeria, even though the former did not participate while the latter did. Some countries joined in simply because they did not want to desert the Arab cause. For Saudi Arabia it was perhaps an attempt to avoid more radical measures whilst at the same time ensuring that the price hikes agreed in Kuwait would in this way be more easily accepted. As a result, the embargo assumed a more politico-symbolic aspect, based on the idea that the major oil companies would in any case supply Rotterdam with enough oil. These various points of view and different objectives made it very difficult for the Den Uyl government, and for Van der Stoel in particular, to take any effective action against the embargo.

3
European Divisions

By the end of October, the situation in the Middle East had still not stabilized. During the night of October 24/25, the Security Council had called on the belligerents to comply with a cease-fire demand and to withdraw to the positions of October 22. A day later Kurt Waldheim, the UN Secretary-General, submitted a plan to station a peace force of 7000 men in the conflict zone for a six-month period. The principal task of this *United Nations Emergency Force* would be to ensure the cease-fire along the Suez Canal, and the withdrawal of all troops behind the lines occupied on October 22 (when the truce should originally have come into operation). The first Egyptian-Israeli talks were held under UN auspices on the 27th.

Despite these developments, a peace accord was still remote. On November 5, 1973, Kissinger left for the Middle East to help work out an Israeli-Egyptian armistice. This effort was rewarded with some success, even if only the agreement of a truce formula for the Israeli-Egyptian front. On November 11 an agreement was signed by the Israeli Major-General Aharon Yariv and his Egyptian counterpart Abdel Ghani Gamasi, under which Israel undertook to withdraw to the positions of October 22. But it quickly became apparent that further details of the truce would be more difficult to work out; and not only were the Israeli-Egyptian negotiations threatened with an impasse, agreement on the Israeli-Syrian front had not even begun.[1]

The war had ended but at the beginning of November, despite all diplomatic efforts, the embargo aimed at The Netherlands went ahead, although not all Arab oil-exporting countries participated. On November 4, the Arab OPEC countries succeeded in closing ranks again with the announcement that November's oil production would be reduced by 25% (compared with September levels). Although this was again a powerful measure, the reductions implemented earlier were discounted, including the embargo in place against The Netherlands, the USA and Denmark (on account of its pro-Israeli stance), so that the Arab resolutions of November 4 appeared worse than they in fact were, as the Dutch Ministry of Economic Affairs observed.[2]

Nonetheless, serious problems were becoming apparent to the consumer countries. As stated by the Dutch Cabinet on several occasions, international cooperation was needed to deal with these developments. Such cooperation should chiefly take shape within the context of the OECD or the EC. In the event, however, it would prove difficult to achieve a common response among the consumer countries, for these countries, even in Western Europe, seemed to be more keenly competitive than cooperative.

Initially, the Dutch government hoped that steps could be taken within the OECD to come to the assistance of countries affected by the embargo: in fact, the OECD had tried in the 1950s and 1960s to develop an emergency allocation system. The Dutch Cabinet also tried to move the European Commission to introduce a proportional sharing of oil within the EC, but it was quickly evident that both these attempts were fruitless. Meanwhile it was learned that the Foreign Ministers of the EC were to meet within the framework of the European Political Cooperation (EPC) to discuss the situation that had arisen. This appeared to be an important conference, for it was hoped that the other EC partners would show solidarity with The Netherlands. The Hague had two objectives: proportional sharing of oil supplies for EC member states, and the maintenance of free traffic within the common market. However, it was a very open question whether the other European partners were prepared to agree on a common policy with regard to the oil crisis: interests within the EC were widely divergent, and in addition, the Arabs were threatening sanctions against countries that lent The Netherlands a helping hand. And the past, too, stood in the way of effective cooperation.

The Netherlands and European Integration

The process of European integration at the time of the oil crisis was in a dynamic phase. The entry of Great Britain, Ireland and Denmark in 1973 had expanded the EC to nine countries. General de Gaulle had for years resisted Britain's entry, but after the General's departure in 1969, agreement had been reached over the Community's expansion. After all the conflicts of the 1960s, this expansion seemed to provide European integration with a new boost, and consequently plans were worked out around 1970 to develop the EC, in good time, into an economic union. In 1971, the six government leaders and heads of state adopted a plan to introduce monetary union in 1980.[3]

This European revival went hand in hand with, and perhaps contributed to, a growing estrangement between America and Europe. At the time there were various conflicts and differences of opinion in play between the West European countries and the USA, not least in the monetary area and over trade. The Nixon government was reproached with its pursuit of an economic and monetary politics all too closely aimed at furthering America's own interests, while at the same time there was increasing criticism in Western Europe and also in Japan of America's prosecution of the war in Vietnam. This irritation in European capitals and in Tokyo was only further augmented by the lack of consultation over such matters as the détente with the Soviet Union, the SALT-I treaty (1972), the approach to China, and the Middle East. Mutual relations seemed to deteriorate so far that in 1973, Kissinger called for the *Year of Europe*, in an attempt to breathe new life into the Atlantic Alliance, albeit on a new foundation.[4] The French historian Alfred Grosser believes, nonetheless, that 1973 was the nadir of postwar American-European relations.[5]

The EC itself had for years been the ground of serious differences of opinion and collisions of interest. The Netherlands had played an active role in these conflicts, not least as the opponent of Gaullist France. The two countries had also opposed each other in the years 1960-63 following France's proposal to reform the EC as a political union. Luns had stood by the principle that the Community must remain an economic community, whose aim was the free economic movement between the Six, in which the institutions created by the Treaty of Rome, and particularly the European Commission, should play a leading role. Furthermore, the Dutch argued, the politicisation of European integration threatened to undermine the unity of NATO.[6] Following the breakthrough of 1969, i.e. the decision to expand the EC to nine members, various plans were mooted in the early 1970s to try to get the political development of the Community moving again. One of these plans, the Davignon report in 1969, advocated more consultation and perhaps even a level of harmonization of foreign policies within the EPC framework. In the second Davignon report, from July 1973, it was proposed that the nine member states, after mutual consultations, should adopt a common standpoint on questions 'in those fields where a common position would be necessary or desirable'.[7] In this same period, the need arose to give more authority to consultations between EC heads of state and government leaders to meet on a more regular basis and even to form an institutionalized European Council, a proposal which would be definitively decided in 1974.[8]

Although de Gaulle had in the meantime disappeared from the stage, the Dutch in the early 1970s remained apprehensive about political cooperation within the EC. The Netherlands opposed the intensification of EPC consultations, nor did The Hague look kindly on the institutionalization of the European Council – though on this point it has to be said that there was a difference of opinion within the Den Uyl Cabinet, between the Premier and his Foreign Minister, Van der Stoel. This difference of opinion was also to play a part during the oil crisis. It should be borne in mind that the institutional changes in the EC were still fresh, such as the introduction of official EPC-preliminary consultations in the form of the *Comité Politique* of the Directors-General for Political Affairs (DGPA's).

In the years leading up to the oil crisis, there had also been clear differences of opinion between France and The Netherlands in the field of energy policy. The Hague objected to the more controlling, *dirigiste*, role of the EC in the provision of energy, and its greater independence from the USA, as proposed by Paris. The Netherlands stood for a free oil market in the EC and for maintaining the existing power structure in the international, and especially in the West European oil sector in which, after all, The Netherlands occupied an important position.

The Dutch government had also resisted plans put forward in 1968 by the European Commission for developing a common policy on the trade in oil and oil products. These were considered at the time as being excessively *dirigiste,* threatening to involve the EC in all kinds of political complications. As we have said, The Netherlands – even after the formation of OPEC – was against direct consultations between the West European countries and the oil producers. As set out in the Explanatory Memorandum for 1972: 'The government commits itself not to get involved in the negotiations between the producer countries and the oil companies.' Unlike those countries with state-owned oil companies like Italy and France, the feeling in The Hague was against any direct state intervention in the negotiations between the oil companies and OPEC, which were seen as 'a purely commercial matter'.[9]

These basic premises were set out again in a note of October 1972, on the occasion of a conference on the EC's energy policy. According to this note, the little progress made in the preceding years toward establishing a common energy policy was mainly due to the very real differences in the energy situation of the different countries of Europe; and above all, the very different perceptions of The Hague and Paris. Paris wanted to promote the assurance of oil provision by obtaining direct influence over imports. In addition, it has to be said, Paris was also working to raise the status of the smaller French oil companies to the same level as Shell and BP.[10]

Such objectives were of course in conflict with Dutch interests. The Dutch Cabinet naturally therefore took an extremely cool stand on the French ideas over market regulation. The Hague was in general opposed to EC intervention in the oil sector. [11] The British entry into the EC was regarded in The Hague as a welcome counterweight to those member states very much in favour of market regulation and state intervention, such as France and Italy; and indeed, British membership was soon making itself felt. Shortly before the oil crisis, proposals were put before the European Commission that were far more in line with the Dutch views, based as they were on the principle of a free, communal energy market.[12] Subsequently, however, as we shall see, the role of the British government during the oil crisis was to disappoint The Hague.

France

In fact, from the very outset, it turned out to be exceptionally difficult during the oil crisis to achieve a common EC approach. There was no question of a common EC policy to the oil crisis itself; on the contrary, the EC proved to be the theatre of conflict and a clash of interests.

Paris had already been busy before the oil crisis enhancing its competitive strength in the oil trade by improving the ports of Marseille, Dunkirk and above all Le Havre. The main objective was to try and strengthen the position of the French state oil companies. These attempts, based on large-scale government support, were at the same time aimed at undermining the position of Rotterdam as the most important centre of the oil trade and the main transit port in Europe. This of course did not go unobserved in The Hague and elsewhere.[13] The Dutch Ambassador in Paris, J.A. de Ranitz, reported that 'quite a number of the French' would not look askance if 'the position of Rotterdam as the first world port [were] put in danger', from which, it was hoped, ports like Dunkirk, Le Havre and Marseille might gain advantage. For this reason, some of the French thought their pro-Arab political stance, certainly in comparison with that of The Netherlands, was beginning to bear fruit. For the time being, thought De Ranitz, 'the French government itself was of the view that a certain degree of solidarity was called for', but despite this, it was quickly apparent in The Hague that French politicians and diplomats in different situations were adopting an attitude that was 'incredibly lacking in solidarity'.[14] In particular, the conduct of the French Foreign Minister, Michel Jobert, was followed in The Hague with rapidly growing dismay and distrust.[15]

During the crisis, characteristic and fundamental differences of outlook between France and The Netherlands were revealed, both over the direction of development of the EC and over the EC's energy policy. In addition to these problems, there appeared to exist similarly antithetical positions on European political and economic cooperation, as well as the American-European relationship. Furthermore, very different conceptions of the conflict in the Middle East also played a role. In fact, all these aspects, both from the French and the Dutch viewpoints, were closely connected. The French president, Georges Pompidou, pursued a policy that might be described as 'Gaullism without de Gaulle', with the qualification that he had exchanged de Gaulle's globalism for a European orientation and the EC.[16] And the Dutch-French clashes under de Gaulle were still fresh in the French memory. According to Grosser, Pompidou complained to a close confidant:

> Les Hollandais detestent la France: c'est la seule constante de leur histoire. Ils nous donnent les leçons sur l'Europe, mais ils ne souhaitent rien d'autre que de l'amarrer à l'Amérique.[17]

And in a similar vein the president told the West German Chancellor, Willy Brandt, that The Hague had never striven for European solidarity. The Netherlands was merely a place of transit: 'un lieu de passage'.[18]

On October 31, De Ranitz reported that Paris did still see something in solidarity. A day later in a meeting with DGPA Van Lynden, the French Ambassador in The Hague, J. Senard, left no possibility of misunderstanding over the French attitude regarding the oil crisis and the embargo against The Netherlands. Senard stated that France was only inclined to work together for a Community solution to the oil crisis if The Netherlands was prepared to accept a Middle East statement in the EPC. And in addition to this, said Senard, The Hague should endorse the principle of a common EC energy policy. According to Senard, it was clear that in times of scarcity the oil companies no longer had an adequate grip on the market mechanism. It was therefore necessary for The Netherlands to adapt to circumstances and to accept that a system of European market regulation had to be established.[19]

In several ways, Italy found itself in a position essentially in agreement with France. Italy also had major state oil companies, although there were differences of opinion on the politics of energy and oil which had become evident shortly before the oil crisis, when reports began to circulate suggesting that Shell wanted to pull out of Italy. Some European-oriented politicians opposed this, since otherwise there would have been only American and state oil companies remaining.[20] Nevertheless, it was quickly apparent that the Italian attitude toward the oil crisis was close to that of the French.

Great Britain

At this stage, the impression began to grow in the Hague that not only Paris but also London was prepared to go to great lengths to secure its own oil supply. This was a blow, given the expectations aroused in The Hague by Britain's entry into the Common Market. London was having to deal with the extraordinary circumstances of a massive miners' strike at the same time as the oil crisis, added to which Prime Minister Edward Heath had to contend with serious misgivings within his Conservative Party as to the wisdom of joining the EC at all. Here then was the opportunity for the Heath government to refute once and for all the assumption that the EC could only be expected to bring misery and misfortune.

Although the differences between the Dutch and the British were less fundamental than with the French, the prospects for 'solidarity' from Britain now appeared equally slight. That, at least, was the message from the Dutch Ambassador in London. On November 1 he reported that the oil companies were being put under pressure 'at the highest level', i.e. by Heath himself,

> to move them to make an exception for the United Kingdom in fixing quotas for Middle Eastern oil, and to maintain unimpaired supplies to the UK; in other words to pass on the reductions applied by the producers to other countries.[21]

This actually happened when Heath summoned Frank McFadzean of Shell and Eric Drake of BP to a meeting as early as October 21. The Shell and BP representatives let it be known on this occasion that because of their role as international players they were unable to comply with this request. But the British government would be able to force them if it came to it, Heath had threatened, by issuing an *Order in Council.*

McFadzean and Drake, however, declined to guarantee a full 100% delivery. When McFadzean remarked delicately that 60% of Royal Dutch Shell was in Dutch hands, an irritated Heath turned to Drake: the British government owned 51% of shares in BP. Drake asked Heath to put his request in writing, to which the enraged Prime Minister replied: 'You know perfectly well that I can't put it in writing'. Drake's response was: 'Then I won't do it'.[22]

The Dutch Foreign Minister, Van der Stoel, wanted to discuss the whole question with his British counterpart, Douglas Home, preferably before the EC meeting of November 5 and 6, but Home had no time, he said. He did write Van der Stoel a personal letter in which he observed that it would be best for everyone if as much oil as possible continued to enter Europe. Communal action co-ordinated by the EC would not contribute to this.[23] This did not sound particularly cooperative, and Van der Stoel instructed Ambassador Gevers to issue a démarche expressing the hope that London would not take 'unilateral definitive steps' before the EC talks 'which could prejudice cooperation within the EC or OECD'. On November 2 Gevers held discussions with the Permanent Under-Secretary at the British Foreign Ministry, D. Greenhill. The Dutch Ambassador indeed expressed the hope that London would take no steps before November 5, which Greenhill was able to promise.[24]

The following morning, Gevers had a further meeting with the Assistant Under-Secretary at the Foreign Ministry, G. Parsons, who said that Arab suspicion was the consequence of the position the Dutch had taken during the political discussions of the Nine, of which 'the Arabs (and Israelis) were always extremely well-informed'. The Arab countries had been given the impression that The Netherlands was the major stumbling block to formulating European support for the Arab cause. Parsons wondered whether The Hague could not stress continued Dutch support for the UN resolution 2949 of December 1972, one of the most pro-Arab resolutions passed by the General Assembly which The Netherlands had supported at the time. According to Parsons, this would give the leaders of the Gulf States in particular room to extricate themselves from the embargo, or at least not to apply it too rigorously.[25]

The Dutch were thus put under pressure from the British side as well as the French to take more account of the Arab standpoints. On November 5, the first day of the EC meeting in Brussels, there occurred an incident in the House of Commons in London which was highly inconvenient to the Dutch: the Labour MP, Christopher Mayhew, said that the British government should be very careful of making any agreement to 'pool oil resources' with countries which, by supplying arms, had *de facto* chosen sides in the October War. The Minister of Trade, Peter Walker, did not react to this accusation, giving the impression that his government attached some credence to such rumours. Gevers asked The Hague whether it was necessary to approach the Foreign Ministry – at least, 'assuming both assertions could be categorically denied'. The Dutch Foreign Ministry apparently had no trouble with this, for the Ambassador was instructed the following day to do just this.[26]

The Neighbouring EC Countries

For various different reasons the conflicts with neighbouring EC member states appeared less acute than with France and Britain. There was more EC solidarity urged from the West German and the Belgian sides than by Paris or London, though neither country was always willing to acknowledge that solidarity in public. Both West Germany and Belgium were much more directly affected by an embargo against The Netherlands because of the importance of Rotterdam for their oil supply. This meant that both countries, to a degree, were in the same boat as The Netherlands. As set out in the Dutch Foreign Ministry report mentioned earlier, 'The Dutch policy during the oil crisis', in his approach toward EC partners, Van der Stoel made 'maximum use of the uncertainty over the question of whether oil destined for transit also fell under the embargo'.[27] And that meant mainly Belgium and West Germany.

This strategy seemed to work. In Bonn there was great uncertainty over the consequences of the embargo, as the Dutch Ambassador F. Kupers reported from Bonn. [28] The embassy in Bonn in fact received many questions about the extent and the consequences of the embargo, especially over the consequences for West German oil imports via The Netherlands. The question was also asked whether the export controls on oil products announced by the Dutch Ministry of Economic Affairs (see chapter 4) also applied to the German emergency stocks prescribed by the EC, which in part were stored in the Botlek area in Rotterdam-Europort and were not formally imported into The Netherlands at all. An arrangement had been made between the two Economics Ministers over these stocks in December 1970, in which it was provided that they in no way whatever fell under Dutch authority, not even in case of crisis. But, as established by Dutch Economic Affairs on more businesslike grounds, the legal form of this agreement was no more than an administrative accord between two ministers. In other words, a new minister need not feel bound by this agreement at all.[29] And in the meantime there *had* been a new minister.

The West German social democratic government under the leadership of Willy Brandt, in the Dutch view, assumed an attitude of far greater solidarity than either Paris or London, even if this was partly through self-interest. Kupers reported from Bonn that high officials of the *Auswärtige Amt* (Foreign Affairs) and the *Bundeskanzleramt* (Cabinet Office) left no doubt that the common market must be maintained. Even the French, it was thought, must realize that economic decline in The Netherlands and West Germany would have serious repercussions on France itself.[30]

In any case, the West German understanding for the Dutch problems in the Arab world did not go unnoticed. The rumours and reports of arms transports over West Germany to Israel also played a role here.[31] The Libyan Ambassador to The Netherlands tried to incite his interlocutors against The Netherlands by declaring that if The Hague could not be forced to adopt a 'neutral standpoint', the Dutch should anticipate a cold winter, which would also endanger West German oil provisions.[32]

Belgium of course was also heavily dependent on oil supplies from Rotterdam. Some 25 million tons of crude oil passed annually through the Rotterdam-Antwerp pipeline, and the Antwerp refineries were totally dependent on this supply. It was therefore of the utmost importance for Brussels to know what exactly the embargo entailed. If it was purely directed against The Netherlands as oil consumer, the consequences for the Belgian market would be negligible, the Dutch embassy in Brussels concluded. But if it was an action taken against the exploitative companies making use of Dutch ports, then the whole question at once became more problematic in view of the fact that Belgium depended on deliveries from these ports.[33]

The Belgian Foreign Ministry took the position that solidarity among the Nine must be the priority.[34] This sounded good, but it did not prevent problems arising between Brussels and The Hague. On October 24, 1973, a ministerial decision was announced in Belgium by which the export of a number of oil products would henceforth be subject to permits. This order also applied to exports to The Netherlands, whereas the converse had until then not been the case in The Netherlands. At the end of October, though, the Belgian policy relaxed a little, i.e. the issue of permits for exports to EC countries would in most cases be automatically granted.[35]

The OECD

The circumstances that had arisen made it desirable for the Western consumer countries to work together to combat the oil crisis, but this was not so self-evident. From the Dutch viewpoint, an obvious framework for the realization of such cooperation appeared to be either the EC or the OECD. The OECD had as an additional advantage the fact that all the main, Western industrialized countries were members, making a European *Alleingang* – to which The Netherlands had always objected – impossible. Furthermore, the matter of emergency measures in times of crisis or of acute oil shortages had been discussed within the OECD over some years. The problems of oil supply during the Suez crisis of 1956 had led the then OEEC (Organisation for European Economic Cooperation, the predecessor of the OECD) to set up a scheme for sharing the burden of a reduced oil supply between the various West European countries. The Middle East war of 1967, when England and France were affected by an oil embargo, had prompted the OECD to develop this scheme further.

The ruling was only applicable to the West European countries. In the 1970s, the situation on the oil market changed rather drastically as a result of rapid growth in the demand for oil and the increasing dependence of Japan and the USA. Washington was rather unwilling to put the vast oil production at home under the common OECD emergency system. In previous years, these developments had led to several fruitless attempts to come to a ruling acceptable to all OECD members. Until just before the oil crisis, ideas over such allocation schemes were still being exchanged within the OECD.

Immediately following the first reports of reduced production, the Dutch Cabinet appealed to the OECD for the enforcement of a proportional sharing of oil if the embargo should go through. The embargo should be considered as a communal matter. On October 25 the OECD Oil Commission met. In a prior EC consultation, the Dutch Permanent Representative made it clear that The Hague advocated the operation of an oil-sharing scheme. If that did not work, the reactivation of the international Advisory Board would be a satisfactory alternative, a group of representatives from the main oil companies whose job it would be to help the Oil Commission to set up the sharing scheme.[36]

Former Secretary-General of the OECD, E. van Lennep, concluded in his memoirs that the OECD allocation system based on unanimity was not put into operation, mainly because of France and Britain.[37] It appears from a report of the OECD meeting that not only these two countries but also the USA played a 'delaying' or 'cautious' role. Most countries did not yet consider the oil situation dire. Reactivating the Advisory Board also encountered resistance. All that was decided was that the chairman of the Oil Commission should ask members of the Advisory Board for information.

There were four reasons for the reserved attitude of most OECD countries; or so it was assumed at the Dutch Foreign Ministry. There was a pervasive lack of information; there was a fear of speculation; there was the worry that ganging up (by the West) could send exactly the wrong signal to the OPEC countries; and finally there was the hope entertained by some countries that they would be treated as special cases. The latter reasoning related to reports that Saudi Arabia considered the oil-consuming countries under three categories: hostile, neutral and friendly. Other considerations played a part in the USA, such as the problem of operating the distribution code.[38] The OECD was thus not mobilized.

The EC

Beside the OECD, The Hague also attempted to get the EC to take measures. Within the EC, too, it was a question of proportional allocation of a reduced oil supply and of organizing this within the commercial operation of the free market of the EC; a point of view which clearly ran counter to the French position. To the Dutch Cabinet, intervention on the part of the European Commission seemed the most desirable option, and in fact the Commission did propose this. At a meeting of Permanent Representatives, the Commission presented a proposal for information exchange concerning oil imports. This was a modest proposal that The Netherlands could live with. However, the Commission further urged on this occasion that proposals mooted in the summer, for a communal EC energy policy, should also be considered.

While the Den Uyl Cabinet, at the end of October, was still hoping for an international solution to the oil problem, a string of Arab countries had meanwhile joined the embargo against The Netherlands. The outlook was hardly a rosy one. During an interdepartmental meeting on October 29 to discuss restrictions on consumption, Economics Minister Lubbers again strongly urged European cooperation. The line of thought at the Ministry for Economic Affairs was that the European Commission would soon have to take measures to curtail oil consumption within the whole EC. If they should fail to do this, then in Lubbers's view The

Netherlands should feel free 'to limit its exports of oil products and natural gas to the European partners'. It was an indication of the assertive attitude held by Economic Affairs throughout the entire crisis. The Dutch Foreign Ministry, however, objected to the threat of retaliation implicit in the point of view of the Ministry for Economic Affairs.

On October 30 the European Commission was asked by The Netherlands to come up with initiatives as quickly as possible so that these could be discussed in the EC consultations of November 5 and 6. The Dutch Permanent Representative in Brussels was asked by his colleagues to press for discussion of the oil problem during the approaching conference. He was instructed at the same time to stress that all countries, when it came to their oil supply, were threatened with involvement. A day later, the Dutch Ambassador in Bonn was also requested to press for the oil problem to be dealt with at the *Auswärtige Amt* during the discussions of November 5 and 6. The situation was, after all, sufficiently serious to justify such consultations. In the first place, then, it was a question of maintaining proportional supplies and at the same time of preserving the common market.[39]

Little was expected in the way of cooperation from London and Paris. There was particularly scepticism in The Hague over French readiness for communal EC action with regard to the oil problem; and any hope of French cooperation was further diminished when it was learned that Pompidou had put out a plan to convene a summit conference of government leaders, to which Heath and Brandt appeared to have pledged their cooperation. It thus seemed that the three largest EC countries were joining forces in a way that held little attraction for The Hague. Furthermore, it was feared that the political aspects of the Middle East conflict would become the main issue during such a summit rather than the oil problem. The Netherlands would then undoubtedly be put under severe pressure. At first, Pompidou's proposal did in fact deal mainly with the discussion of the European attitude toward the Middle East. Only later did the French agree to address the oil question.

During a meeting with the EC Ambassadors on October 29, Van der Stoel once again argued passionately for European cooperation in the oil crisis. He pointed to the important transit function of Rotterdam and warned of the consequences of a *sauve-qui-peut* devil take the hindmost mentality. For the time being, however, his message still fell on deaf ears. Otherwise, The Hague was taking good care that these efforts at European cooperation did not get in the way of relations with the USA. The very same day, the State Department in Washington was informed by Ambassador Van Lynden of the meeting with the EC ambasadors. Van

Lynden emphasized that the Dutch effort to get some form of EC action should not be interpreted as implying any loss of interest in possible arrangements through the OECD. Washington evidently took it for granted that the Dutch should for the present be capable of solving their own problems within the EC. But that was not now the intention.[40]

On October 31 the European Commission made known a number of draft decisions concerning commercial trade in crude oil and oil products. These proposals came down to a procedure over the provision of information to the Commission over oil stocks and oil imports from non-EC countries, and additionally a duty to report exports to non-EC countries. And finally, the Commission proposed draft regulations relating to trade within the EC, based on the assumption that existing trade channels should be maintained. It was thought that member states would continue to issue licences automatically; but the Commission was given authority to suspend temporarily the obligation to issue export licences in cases where this would seriously endanger supplies in a member state.

These proposals met with approval in the Dutch Foreign Ministry, though there were doubts as to whether Paris, Rome and Brussels would agree. For tactical reasons it therefore seemed best for The Hague not to expose itself too much. Current stocks and needs besides made it unnecessary to deviate from normal, automatic issuance of licences. It was more a matter of making an arrangement such that the issue of licences could be halted if the circumstances so required.[41]

In a sub-committee of the Dutch Council of Ministers, the Ministerial Council for European Affairs, Lubbers also expressed his satisfaction on November 1 over the speed of this action taken by the Commission. The embargo was now at least seen as a communal problem. But he was not entirely satisfied. Economic Affairs wanted to tackle not just the provision, but also the use of oil. The rationing of consumption should also be dealt with at the communal level, yet there was no proposal along these lines. Lubbers tended toward the holding of bilateral talks with his foreign colleagues, but that did not go down well with Van der Stoel, who warned against holding 'premature talks'. The first step should be a meeting of Foreign Ministers in Brussels, said Van der Stoel, thus underlining that it was primarily within the competence of the Foreign Ministry to seek a solution to the international oil crisis.[42]

The Dutch Permanent Representative at the EC and the Ambassadors in other member countries were then instructed to act as discretely as possible in order to prevail on especially the French and the British to cooperate fully. Meanwhile it had become clear that the Commission's proposals, at France's insistence, would be dealt with highly confidentially during the coming EC talks, possibly during a dinner for the Foreign Ministers. Discretion was therefore of the utmost importance.

There were nevertheless still objections within the Ministry for Economic Affairs to the Commission's proposals. As a result, the old differences of opinion between Economic Affairs and the Foreign Ministry surfaced anew. As observed in a memo from the Head of the Department for Economic Cooperation (DES) of the Foreign Ministry to Van der Stoel, the proposals provided that export restrictions to member states would only be permitted after it had been established that oil supplies in the exporting country were 'seriously' endangered. This, according to the Head of DES, meant that 'The Netherlands would allow a weapon to slip from its hands to which the Ministry for Economic Affairs was rather attached', viz. the possibility of introducing, or threatening to introduce, restrictions on export or transit goods. After consultation with Economic Affairs, therefore, the permanent representative at the EC was asked to try to get the word 'seriously' removed. This would leave more possibilities open for exceptions to the automatic issue of licences. A second problem was the provision of information that the Commission proposed. Economic Affairs and Shell had up till then been handling industrial information confidentially, since this information could be used as the foundation for a community policy regarding oil supplies that would move too far in the French and Italian direction. This aspect, according to DES, seemed meanwhile less problematic. Shell appeared to change its position on this point.[43]

There were more objections from Economic Affairs. As mentioned above, the Commission also wanted to restart consultations over a common EC energy policy. Economic Affairs was not inclined, however, to cooperate on a Community market arrangement as long as it was not clear what goals such an arrangement would serve. Within the Foreign Ministry the problem was viewed very differently. In a note from the Head of the Department for European Integration (DIE), it was stressed that France was only willing to cooperate on finding a solution to the current oil crisis, whereas The Netherlands was ready to work out a Middle East declaration under the aegis of the EPC as well as cooperate on setting up a Community energy policy. In view of the fact that the major oil companies no longer had a firm grip on the situation on the international oil market, the Committee argued, it was perhaps inevitable that the Netherlands would have to move with events and adapt to the new situation.[44] Contrary to the view in Economic Affairs, that The Netherlands' partners in the EC should come to some arrangement that would leave the import of oil by the Nine to the oil companies, DIE did not want to be isolated within the EC and was ready to put aside the usual Dutch principles in order to get discussions going there.

A Declaration by the Nine

In the meantime, several countries were arguing that the EC talks fixed for November 5 and 6 should be devoted mainly to discussion of the Middle East situation. Most member states wanted to concentrate on the *political* aspects of the war, not in the first place about the consequences of the Arab oil embargo. The Hague, however, was mainly interested in EC cooperation with regard to the oil crisis, not in a common Middle East policy. However, once it was settled that the oil problem would also come up for debate, Van der Stoel felt able to agree with the proposal to use the talks to discuss the war in the Middle East as well.[45]

Yet in the event, The Netherlands' partners paid far more attention to the political side of the question, and consequently to the differences of opinion within the EC on this issue. And in this context, on November 1, the British again asked the Dutch Cabinet whether they would not bring Arab attention to the fact that the current Dutch government, just like its predecessor, endorsed the UN resolution 2949 of December 1972. In this resolution, rejected by Israel and the USA, the rights of the Palestinians were recognized as an essential part of any peace settlement in the Middle East. Van der Stoel, who as a member of the Dutch Parliament had once criticized the Dutch support for this resolution, declined the British suggestion.[46]

In spite of all this, in early November a new draft government statement was prepared at the Dutch Foreign Ministry, which referred to the PvdA position that the Palestinians had the right to their own political identity. According to this draft, the government should state that it shared this judgement. Largely because of the reference to a party political position, the draft was quickly put aside, and a new instruction drawn up for the Ambassadors in the Arab countries which was approved by Van der Stoel on November 5. It was observed in this document that the Arab countries were evidently pressing for further enlightenment on The Hague's position over future Israeli borders and the Palestinian question. With regard to the first point it was said that the Israeli borders, as stated earlier, must be approximately the same as those existing before the 1967 war. The question of whether The Netherlands adhered to the French or the British version of resolution 242 was not raised. On the Palestinian question it was specified that their political aspirations had to be embodied in some form or other as an essential part of any future peace settlement. This signified a definite sharpening of the Dutch point of view, even though nothing had been said about a Palestinian political identity, let alone a Palestinian state.[47]

Meanwhile, the long-awaited EC ministerial conference was approaching. The week from October 30 saw the preparations for reaching a common political standpoint in the *Comité Politique*. Progress was difficult, partly through fear of leaks to the press, as was later to happen during the ministerial talks of November 5 and 6. In a first draft text the Nine announced that they wished to play an active role in ending the Middle East conflict, within the framework of the United Nations, and specifically on the basis of resolution 242. But on November 4, the first car-free Sunday in The Netherlands (see Chapter 4), the British DGPA submitted an entirely new draft statement.

On the Dutch side there was a major objection to a passage proposed by the British in which the Nine reasserted their support for resolution 2949 and subsequently listed a whole series of ingredients that would have to be part of any peace settlement: 'the non-acquisition of territories by force', the 'withdrawal of Israeli armed forced from occupied territories', the 'respect for the sovereignty, territorial integrity and independence of every state in the area' and 'the invalidity of changes carried out by Israel in the occupied territories'. The British proposal also included a recognition of the rights of the Palestinians.

It appeared in the *Comité Politique* that The Hague objected only to this passage. At the suggestion of the Belgians, the reference to 2949 was removed: a list of ingredients essential to a peace settlement would have to be sufficient. But agreement on this list turned out to be not such a simple matter, in particular the withdrawal of forces from the occupied territories.[48] From the final closing statement it seems that DGPA Van Lynden successfully resisted the passage concerning the changes to be implemented by Israel in the occupied territories. Furthermore, a passage was added on the right to secure recognized borders, undoubtedly for the benefit of Israel.

Finally, Van Lynden, who had been in constant contact with the Dutch Foreign Ministry, approved the draft declaration. The Netherlands thus adopted a position of loyalty, as was also later and emphatically repeated, even if this approval was granted in the hope that an EPC declaration could be 'linked' to a common EC action in response to the oil crisis. The latter was referred to the ministerial discussions, which was not a good sign.

On Monday, November 5, Van der Stoel arrived in Brussels. The Benelux ministers held a meeting that afternoon and arranged to keep in contact with each other. On Monday evening, Van der Stoel spoke for an hour with his British counterpart, Douglas Home, and the British Energy Minister, Ralph Davies. It was clear in the course of this meeting that London was strongly against any EC measures relating to the trade in oil and oil products. Britain and France refused to attend to the Commission's proposals. Home declared that Britain was still receiving a large quantity of Arab oil and that his government did not want to give the Arab countries any pretext for stopping it. The British fear that the Arab countries could exercise control over the transfer of the flow of crude oil was not shared by The Hague. From the Dutch side, it was pointed out that, according to Shell, it would be difficult to control whether in the coming period more oil entered Rotterdam from Iran and Nigeria and less into British ports, but the British stuck to their viewpoint.[49]

It was during this conversation, according to Grünfeld, that Douglas Home threatened that London would not hesitate in an emergency to use legal means to compel the oil companies to keep to their contracts.[50] Perhaps Home said this in reaction to a statement by Shell's president, Wagner, who had the previous day publicly stated that the oil companies would share oil shortages in Europe between the various consumer countries. In spite of this, Van der Stoel put it to the Dutch Council of Ministers three days later that the British government would not obstruct mutual agreements made between the major oil companies. Then PvdA leader Van Thijn also noted in his diary that Douglas Home had said: 'we will not interfere with the commercial policy of the oil companies'.[51]

The British Minister Davies had warned Van der Stoel about entertaining illusions concerning the role of the oil companies. Davies could see no way that influence could – or would – be exerted on the oil companies to move crude oil, despite the embargo, to Rotterdam; though with refined products it was a different matter. Van der Stoel's argument of the threat of disintegration of the EC did not convince his British colleague. 'Even my plea to keep strictly secret any measures that might be necessary – such as those proposed by the Commission – made little impression', Van der Stoel later reported. Home held the view that there was absolutely no possibility of secrecy in Brussels.[52]

The fear of leaks also turned out to play a major role later, in the dinner attended by the nine Foreign Ministers and the Chair of the Commission, François-Xavier Ortoli. According to Van der Stoel, it transpired that the fear of any common action drawing to itself the attention of the Arab countries was widely shared. Only those countries that depended on the movement of oil through Rotterdam declared themselves to some extent prepared to take joint action. The French Foreign Minister Jobert, above all, would not hear of any joint action, referring to the fact that the Commission proposals had meanwhile been published in the French press. Jobert did not want to discuss these proposals over dinner.

The Ministers quickly agreed on the prepared EPC declaration over the situation in the Middle East. The communiqué of the Nine stated that the EC wished to play a role in finding a solution to the Middle East conflict, effectively going against the Americans' virtual monopoly, until then tacitly accepted, of Western intervention in the Middle East. The Nine expressed the hope that negotiations could be opened under the aegis of the UN, and based on the implementation of all sections of resolution 242. A peace accord must be based on four fundamental principles: 'the inadmissibility of territorial expansion through violence'; 'an end to the territorial occupation of areas controlled by Israel since 1967; 'respect for the sovereignty, integrity and independence of all states in the region, as well as the right to live in peace within secure and recognized borders'; and finally 'the legitimate rights of the Palestinians'.

The Dutch acceptance of the second point appeared to be a volte-face. Van der Stoel subsequently said that he had made an interpretative statement on November 6 in which he had said that The Hague construed this second principle as conforming with resolution 242. 'In any case, this back-up statement without doubt lays down that The Netherlands had not committed itself to a formal evacuation of all occupied areas.' He also remarked here that making this information public would have perhaps made his position in regard to domestic politics rather more comfortable; but he had not done this because it would not have looked so sensible in regard to the oil countries.[53]

Neither during the EPC meeting nor in a subsequent closed session, where occasionally 'frank words were exchanged', were any new viewpoints on the oil crisis raised. Only the West German Foreign Minister, Walter Scheel, called for solidarity with partners who found themselves in diffculties. Van der Stoel too urged cooperation and, in view of the Britsh fears on this score, as much secrecy as possible, but he got absolutely no support from either the French or the British sides. Jobert stated in the clearest terms that the Dutch Middle Eastern politics had led to the embargo against Rotterdam and therefore no appeal to the Common Market was justified.

There was thus no support for The Netherlands, despite the fact that the majority public opinion in the EC was for help to member states which got into difficulties (varying from 59% in England to 70% in France and 88% in Italy).[54] London seemed only interested in securing its own oil supplies. Brussels, Luxembourg and Bonn showed some understanding of the Dutch position, but this understanding would go no further than verbal support. Luxembourg and Belgium pushed for a press statement on the whole oil problem, apart from anything else to avoid the EC losing credibility.

The press statement was duly issued and spoke of the mutual interdependence of member states' economies and of the necessity to follow the situation closely. To Van der Stoel's exasperation, however, it announced not a single concrete measure to be taken. He had, after all, agreed to a new Middle East declaration on the assumption that actual cooperation in the matter of oil supplies would also be addressed. That was evidently a mistaken assumption. The Dutch Foreign Minister wondered whether disintegration of the EC could in the end be avoided.[55]

Reactions in The Netherlands

As indicated earlier, the policy pursued by Van der Stoel was not without its critics within the Dutch Foreign Ministry. DGPA Van Lynden and, as we saw earlier, the Department for European Integration (DIE) were also of the view that The Netherlands should avoid becoming too isolated within the EC. This difference of outlook also involved the general question of whether The Netherlands had to accept the EPC as a European policy-determining framework more than previously. In both these respects, accepting the statement of November 6 meant a change of policy, or at least a first step in that direction. It meant acceptance of the EPC as a policy-shaping and policy-co-ordinating executive body, and a provisional end to The Hague's isolation within that body. It also meant an accommodation, albeit a cautious one, to the British and French positions, though the significance of this should not be exaggerated, given Van der Stoel's 'interpretative statement' cited above. The passage concerning the occupied territories was vague (any reference to *the* occupied territories was avoided), and The Netherlands had earlier expressed recognition of the legitimate rights of the Palestinians by accepting resolution 2949 in the UN General Assembly.

At first, at least, the EPC statement of November 6 was not well received. According to oil expert and author Daniel Yergin, Arab diplomats considered it 'a kiss blown from afar – which is all very nice, but we would prefer something warmer and closer'.[56] Reactions in Israel, on the other hand, were bitter: it was felt that the statement was a step in the direction of the Arabs. Kissinger was also dismissive, speaking of a 'stampede of dissociation' and complaining of a lack of consultation on the part of the EC. He announced that his country 'was going to pursue its own policies in the Middle East in any event'.[57] Relations between the USA and America's European NATO partners did not appear to have improved; and because of this Brandt declared a week after the signing that the Nine had gone too far.[58]

In The Netherlands, too, there was much criticism of the declaration. The progressive newspaper *de Volkskrant* concluded that the November 6 communiqué made far fewer compromises to the Israeli position than the American peace plans. The declaration therefore had to be seen as a French diplomatic success.[59] *Het Parool* (social-democratic) also regretted that The Netherlands had signed the declaration, for it had evidently been the fruit of Arab pressure and because the Nine had been made 'the instrument of the French Middle East policy'.[60] *Trouw* (protestant) spoke of a 'bourgeois timidity hiding behind the oil stove'.[61]

Of all the influential Dutch newspapers, only NRC *Handelsblad* found little to object to, taking a rather laconic view. It was a question of a declaration from countries not directly involved in the conflict who did not want to engage in world politics, but merely wanted to defend their own regional interests. The declaration contained 'nothing more than opinions, without indicating the means by which the proposed goals might be achieved'.[62] This was neither the first nor last time that NRC *Handelsblad* had tended to give prime importance to the interests of the port of Rotterdam. Three days later the paper published an open letter from the historian L. de Jong, roundly accusing Van der Stoel of having betrayed Israel and thus ending the previous solidarity with that country, which De Jong referred to as a 'debt of honour'. NRC *Handelsblad*, in its editorial column, again reacted in pragmatic fashion. The paper called the EC declaration 'healthily realistic', and it did not necessarily mean that Israel had been repudiated by The Netherlands. It was merely a question of giving a 'slight change of tack' to Middle Eastern policy.[63] Reproaches were also voiced in the Dutch Council of Ministers, notably from Vredeling, who regretted that the EC had not condemned the use of the embargo as a weapon. By yielding to an embargo, one only put oneself in a more vulnerable position which, according to him, could also be to Israel's cost. He would have preferred to distance himself from the declaration. He regretted the British attitude and had let this be known in no uncertain terms in NATO's Nuclear Planning Group, which his British counterpart initially did not want raised for discussion, although it had eventually happened. In the future, The Netherlands should do everything possible to prevent the French line being followed. He again referred to the fact that Israel was not fighting on its own strength, but with resources of 'mainly' American origin. All in all, Vredeling's was a bluntly Atlantic-oriented argument. The government had to put its trust in the USA. The American delegation had informed Vredeling that The Netherlands would not be left without oil.[64]

The Dutch Minister of Finance, W. Duisenberg, on the other hand, had only praise for Van der Stoel, whom he thought the target of a witch-hunt. Lubbers too supported Van der Stoel and emphasized that the business world thoroughly approved the consistent line taken by the Cabinet, not to submit to threats, even though Dutch economic interests in the Arab region were great. Den Uyl also thought it prudent that The Netherlands had not distanced itself from the other eight, not least in connection with the embargo. Further, the limits of what was still acceptable to Israel had been properly taken into account in the declaration of the Nine. He again cited the fact that Van der Stoel had had to work with the fiercely pro-Israeli sentiment in the Second Chamber which, after the embargo, had suddenly been reversed.[65] And Den Uyl was right.

Besides, before arriving at any judgement of the signing of the new EC declaration, it is important to point out that the character of the war in the Middle East had changed entirely. Israel's position had in the meantime so improved that it was no longer endangered, however perilous those first days of the war had been. For this reason alone it would be a great oversimplification to posit that the EC, and The Netherlands too, had undergone a complete volte-face because of Arab pressure.[66]

Conclusion

The embargo against The Netherlands went ahead. It was particularly difficult to raise international support for The Netherlands' problems. On the contrary, it appeared that for the time being Britain and France were only too willing to make the most of the difficulties in which the Dutch were embroiled. Both within the OECD and the EC, it proved impossible to arrive at any common response to the oil crisis. Even the readiness of The Hague to endorse the new EC declaration on the Middle East had not been able to elicit support for the Dutch. In the EC, it was rather a bruising affair with no holds barred; and for that matter, the Cabinet also joined in. As we saw earlier, initiatives like the De Lavallette mission (see Chapter 2) were taken outside the EC. Furthermore, Lubbers threatened openly to turn off the natural gas tap, and Van der Stoel's policy was partly based on creating uncertainty in those countries which depended on imports from The Netherlands.

But so far, all this manoeuvring within the EC produced little result. The Netherlands' EC partners, it seemed, were not inclined to a sense of 'solidarity'. It is nevertheless the question of what concrete support could have been expected from those EC partners. There was a considerable difference of opinion on this point between the Dutch ministries for Economic Affairs and Foreign Affairs. Economic Affairs mainly wanted talks to deal with concrete problems: both the supply problem and the necessity for restrictions on consumption. Should there prove to be no readiness for cooperation among the EC partners, argued Economic Affairs and Lubbers, a more clinically businesslike approach would be necessitated that would not attempt to disguise the existing conflicts of interest within the EC. The Foreign Ministry was in general more cautious and wanted to prevent such conflicts of interest being forced into the open. Foreign Affairs even went so far that it was ready to discuss – in the context of the EC – a common European energy policy if the partners, meaning specifically France, would agree to take measures to support The Netherlands if hit by an embargo.

Economic Affairs took a much more cautious view of this last proposal, and here the long-standing conflict of competence in the international arena of the oil problem undoubtedly played a role. Yet it was Economic Affairs that took the more cynical view of European cooperation than did the Foreign Ministry. Secretary-General of Economic Affairs, F.W. Rutten, later said he thought that all the pleas for EC solidarity were intended purely for public opinion. In his view, such arguments were regarded by those involved in Economic Affairs at the time with considerable scepticism.[67]

Meanwhile, it was clear that the oil companies wanted to divide equally the difficulties caused in Europe, including the embargo. The Netherlands was therefore not in quite such a bad position as initially thought. The Cabinet would, in fact, review its standpoint on 'solidarity' during the course of the following November; but the Den Uyl Cabinet was still confronted with Arab cut-backs on production and the threat of reduced oil supplies. This meant that The Netherlands, like other consumer countries, had to ready itself for rationing the domestic use of oil and oil products. In the following chapter, we shall therefore turn our attention to the wide range of domestic measures taken by the Den Uyl Cabinet in order to reduce the consumption of oil and oil products.

4
Domestic Measures

In October and November of 1973, The Netherlands was confronted with a series of threatening Arab moves. Following the drastic rise in the price of oil on October 16, came the decision of the Arab OPEC states a day later to reduce oil production by 5% each month as long as the Western countries continued to support Israel. Almost a week later, The Hague was confronted with a full embargo, even though in the end not all the Arab oil states joined in. On October 18 Saudi Arabia itself announced that oil production would not shrink by 5% but by 10% monthly until all the Arab demands had been met with. On November 4 the situation appeared to worsen even further when the OAPEC countries took the decision to reduce production by 25% from September levels. By this time, it began to be clear that The Netherlands stood alone.

The Den Uyl Cabinet

The Den Uyl government had been in power in The Netherlands since May 1973; a coalition consisting on the one hand of the three 'progressive' parties, the Dutch Labour Party, the *Partij van de Arbeid* (PvdA), the progressive-liberal *D'66* and the radical *Politieke Partij Radikalen* (PPR), and on the other hand, the Christian Democratic parties, the Roman Catholic *Katholieke Volkspartij* (KVP) and the Protestant *Anti-Revolutionaire Partij* (ARP). After lengthy and difficult negotiations this coalition was put together in the spring of 1973. With the exception of the short-lived Cals Cabinet, the PvdA had not participated in government since the collapse of the fourth Drees Cabinet in 1958. The PvdA seemed to play a central role in the Den Uyl Cabinet: apart from the premiership, PvdA party members occupied ministerial positions at Foreign Affairs,

Defence, Finance and Development Cooperation. Furthermore, the three progressive parties, the PvdA, D'66 and the PPR, together held a majority of seats in the government.

The Den Uyl Cabinet would seem to have been a product of the social upheaval of the 1960s. At the elections of late 1972, the PvdA, D'66 and the PPR had presented the electorate with a common social programme, entitled *Turning Point '72*, which promised structural reforms in several areas, both in domestic and foreign politics. This programme even spoke of reducing the might of big business.

Turning Point '72 was of course no policy for government. The Cabinet was a coalition, and furthermore the Christian Democratic parties with support from the right had a majority in the Second Chamber. Nevertheless, Den Uyl announced in May that his Cabinet would try to contribute to a fairer distribution of knowledge, incomes, wealth and power, by means that included adapting the system of taxation. In the autumn of 1973, various measures were announced to stimulate the economy and improve the employment situation through increased government spending.

Unsurprisingly, therefore, the Den Uyl Cabinet seemed to have no obvious allies among employers and big business with whom to discuss policy in times of serious political or economic crisis. Throughout the oil crisis, blame and criticism were to be heard that either directly or indirectly referred to the progressive character of the government – for having adopted a too frivolous, irresponsible position, of having paid too little attention to the interests and the viewpoint of employers. Van der Stoel was also accused of being moralistic, of having unrealistic pretensions that could actually damage real economic interests. In various ways, the oil crisis was a first major *test case* for the Cabinet.

The Importance of Oil

There can be no doubt that the steps taken by the Arab OPEC countries were a serious threat to the Dutch economy, and in addition a major setback for the Dutch government's aim to stimulate the economy and achieve a more egalitarian society. The announced price rises, the restricted production and of course the embargo all contributed to this threat. The newspaper NRC *Handelsblad* predicted on October 20 that the dearer, scarcer oil would lead to a decline in affluence.[1] 'Turning off the oil tap', pronounced *Het Parool* few days later, was on paper no less than a national disaster.[2]

It was clear that the embargo could have far-reaching consequences. It had been calculated for Den Uyl at the end of October that The Netherlands was dependent on oil for roughly 53% of its total energy needs. Leaving aside the effects of the price rise announced on October 16, some two-thirds of this oil came from the Middle East. This meant that around 40% of The Netherlands' oil supply was now in the balance.[3]

On the other side of the scales, 45% of The Netherlands' energy needs were satisfied from their own natural gas reserves. This greatly reduced the country's vulnerability and its dependence on oil imports, which were correspondingly less than many other EC member states. The UK and West Germany with their coal and The Netherlands with its natural gas, were in fact less dependent on oil for their energy needs than France or Italy.[4] Indeed, The Netherlands could be considered 'the most energy-rich country in Western Europe'; and so it was considered at the time by the Dutch Ministry of Economics.[5] Considerable quantities of gas were exported to West Germany, Belgium and France. In 1973, around 58 million cubic metres of natural gas were produced, a figure which rose to 70 million in 1974.[6]

Beside natural gas, of course, the Dutch also exported refined oil products. In 1972, the total throughput of oil involving all Dutch ports was around 149 million tons, of which 81 million tons were forwarded by ship, lorries or pipeline (West Germany being the greatest purchaser). This left a net import into The Netherlands of 68 million tons.[7]

However, oil was also an important raw material for the Dutch oil processing and petrochemical industry. The Arab oil measures could therefore have enormous consequences and could well lead to inflation and undermine prosperity, although exactly how this would come about was, of course, difficult to tell. The dependence on oil, and the concomitant susceptibility to price rises, restricted production, and the embargo itself varied widely between different economic sectors. Road traffic and road transport, so important for the Dutch economy, were almost totally dependent on oil; and the same was true for shipping, air transport, the fishing industry and the greenhouse horticultural industry.

But in fact, oil imports were not only important as a source of energy for the Dutch nation; the significance of Rotterdam and the Rijnmond area lay for a large part in the supply, processing and transit of oil. In the Rijnmond area alone, some 20,000 people were directly dependent on the influx of oil and its processing. According to first estimates, the embargo affected approximately 70% of all the oil arriving in Rotterdam. Refining and chemical industries in the Rijnmond area were also completely dependent on the supply of crude oil. More generally, oil was an important industrial raw material, particularly for the chemical industry.

The Botlek or Rijnmond area (that area of Holland where the Rhine flows via its many mouths into the North Sea) and Rotterdam formed a crucial link in the provision of oil to Western Europe as a whole. In 1972, about 70% of the European oil supply came from OAPEC countries, about 25% of which was delivered to Rotterdam. The Rijnmond area was also hugely important for oil processing. The refineries there, comprising some 10% of Europe's refining capacity, exported approximately 75% of their product. Rotterdam was thus an extremely important centre of the oil trade, immensely significant for the whole of Northwest Europe, and furthermore one of the principal refuelling ports in the world.[8]

Depending on the point of view, the central role of Rotterdam and the Rijnmond area in the oil sector could be seen either as a weakness or as a strength. The dependence on oil deliveries would at first sight suggest vulnerability, wrote NRC *Handelsblad,* which continued to speak as a powerful champion of the interests of the Rotterdam port throughout the entire crisis: the embargo 'could be disastrous for our image'. There was an 'enormous danger' that Rotterdam's dominant position as a port would be affected. The paper pointed out that the business climate for major enterprises in The Netherlands had deteriorated over recent years. If the image of The Netherlands now gained ground as a country with difficulties in the oil sector, this could be fatal for Rotterdam's position as the largest oil port in the world.[9] *De Volkskrant* also feared a 'permanent loss of Rotterdam's position' as a consequence of the Arab actions.[10] But Rotterdam's crucial role at the same time allowed the possibility of restricting transit, not only in the interest of Dutch economic needs but also as a way of putting political pressure on surrounding countries. In discussions held on October 30 between the Rotterdam City Council and the Dutch Ministry of Economics, there was speculation over the possibility of curtailing the free transit of oil.[11] Natural gas was another candidate for restricting exports and applying political pressure.

Uncertainty

Initially, there was uncertainty over what the actual consequences of the OAPEC countries' actions would be for the oil supply. In particular, because the sailing time for oil tankers from the Persian Gulf via the Cape of Good Hope to Rotterdam was about six weeks, it would be the end of

November before it became clear exactly what the effects of the embargo and restricted production would be.

Clearly, it was of overriding importance to obtain accurate information. On October 16, the same day that Kuwait decided on substantial price increases, a letter was sent by the Deputy Director for Coal and Natural Gas at the Dutch Ministry of Economics, H.A.A.M. van Eupen, to the oil companies, with the message that the current arrangements for mutual exchange of information must be stepped up, and the usual monthly supply of data on deliveries, production and stocks increased.[12] Indeed, there was a general intensification of consultations between the Ministry of Economics and the oil companies. The former director of Dutch Shell said later that the oil companies had had 'the closest contact' with the Ministry 'on a virtually daily basis',[13] usually with the Directorate-General for Energy.[14] Connections with other agencies and organisations concerned were also strengthened by the Ministry of Economics. Thus, on October 19 the first meeting over the energy shortage took place with the gas and electricity companies and with the oil-producing and distribution companies.

The Dutch Council of Ministers of October 19 was the first occasion for the exchange of ideas involving the entire Cabinet on domestic measures for rationing the use of oil. Lubbers had in the meantime suggested to the Second Chamber that a ban on driving would be a simple solution to restrict fuel use; but Den Uyl, urging restraint, argued that an appeal to exercise economies would be sufficient for the time being. Lubbers too recognised that the first stage in any programme of measures should consist of information. There seems to have been agreement that, as long as it remained unclear how grave the situation was, an informational campaign together with an appeal for everyone to economise would be sufficient. In view of the currently prevailing uncertainty, the Council of Ministers decided that Lubbers should use his own discretion.[15]

During a staff discussion at the Ministry of Economics several days later, it was in fact concluded that so far there was no question of any cutback in oil deliveries to Rotterdam. Although Algeria had meanwhile announced an embargo, the share of Algerian oil in the total supply was no more than 1 or 2%. Most of those present at these discussions felt that as long as there was no concrete evidence of a slow-down in the flow of oil, no compulsory measures to curb consumption should be introduced. Furthermore, domestic use in normal times amounted to only 40% of the supply. If measures did have to be introduced, for the time being Lubbers preferred that they should be limited to voluntary restrictions. He was therefore prepared to request the oil companies to reduce deliveries to the distributors and to launch a campaign to encourage a reduction in oil use.[16]

This, however, according to the Ministry staff, should not disguise the fact the Dutch government had to adjust to harder times ahead. Indeed, the first steps in this direction were taken on October 22 when it was decided that the oil companies would be required to submit data weekly rather than monthly on supply, stocks, processing and export. It was further decided to subject the export of the more important oil products to licensing on the basis of the Import/Export Law, whereby the issue of licences for export to EC countries was automatic. For non-EC countries, it was a precondition for the issue of the licence that there must be no reduction in the stocks held by The Netherlands. This licence system would also apply to the other Benelux countries, though this was not a judgement immediately adopted by the Cabinet.[17]

As already said, it was at that time still difficult to say in real terms what the consequences of the Arab actions would be. According to some reports, there was in reality no embargo in actual operation (as we saw in Chapter 2). Reports were reaching the Dutch Foreign Ministry that once tankers had left Arab ports 'no further control could be exercised over their destination'.[18] Elsewhere in Western Europe, there was similarly no evidence of any reduction in the flow of oil.[19]

Despite this, the Cabinet's policy of restraint was challenged, even within the Ministry of Economic Affairs. The political scientists Rosenthal and Scholten believe that there were probably differences of opinion within Economic Affairs over the need for quick action; specifically, disagreement between the Directorate-General for Energy (DGE) and the Directorate-General for Industry (DGI). The Directorate-General for Energy advocated immediate, sweeping, across the board measures, whereas the DGI, in view of the major importance of especially the petrochemical industry, argued for measures that would be in the first place voluntary and focused on private consumption.[20] F.W. Rutten, at the time Secretary-General of the Ministry of Economic Affairs, sees these political differences as merely reflecting differences of responsibility. According to several participants in this discussion, the DGE point of view prevailed.[21] Yet it was mainly the uncertainty of the situation that dictated a certain caution. Moreover, within the Cabinet there was no agreement on the necessity for any restrictive measures.

The press lost no time in censuring the Cabinet for its rather laconic attitude. *Trouw,* on the assumption that arrangements were already in place for introducing car-free Sundays and speed restrictions, wrote that there was no reason for not implementing them at once. Such economies could, it was argued, turn out to have other benefits. Furthermore, speed restrictions would help reduce road accidents.[22] NRC *Handelsblad* also reproached the Cabinet for its rather spineless attitude toward the embargo. In the view of one editorial writer, the government was simply not prepared to make an urgent appeal to the Dutch people to limit petrol consumption as much as possible.

'It would be to the Cabinet's credit to show a greater awareness of the problems created by the identification of The Netherlands with Israel in the eyes of the world, and a more vigorous approach to the consequences of this situation', the paper said.[23] Over the following days, the message was repeated by NRC *Handelsblad* that the Cabinet seemed not to be taking the Arab threat seriously enough. Other dailies had much the same criticism: *de Volkskrant* thought that the government had for too long shut its eyes to the situation.[24]

Reducing Consumption

By the end of October it was becoming clear that The Netherlands was threatened with an exceptional position. Although it was uncertain just how effective the embargo would be, this position could seriously damage the economy. In a meeting held on October 30 with Lubbers at the Ministry of Economics, a delegation from the Rotterdam city council reckoned that the port was faced with considerable loss of income and that employment would be endangered. On this occasion, different methods of combating the effects of the embargo were discussed, including the reverse pumping of oil through the pipeline between Rotterdam and Antwerp. Restricting deliveries to surrounding EC countries was also discussed; although to those present the possibility of actually carrying this out seemed small. As the Alderman J. Riezenkamp emphasized, it was crucial for Rotterdam to preserve its relationship of trust with its trading partners. Lubbers concluded on this occasion that everything possible must be attempted to reach the point where the burden would be shared internationally.[25]

Lubbers' view of *burden sharing*, however, was rather optimistic. It was established during a staff discussion at the Ministry of Economic Affairs that the OECD was for the time being not prepared to put into operation any scheme for emergency oil provision. Nor was it very plausible that The Netherlands' EC partners would be prepared to switch to joint actions. Meanwhile, the oil companies, at least as far as Europe was concerned, introduced a general reduction in the oil supply of around 15%: a substantial percentage. The staff discussion group considered that this meant a necessary reduction in domestic use of about 10%.[26]

In order to achieve this 10% reduction, it was decided at Economic Affairs, most probably on October 29, to bring in the Rationing Law and to activate the *National Bureau for Oil Products* (*Rijksbureau voor Aardolieproducten,* RBA).[27] Certainly no modern political instrument, the Rationing Law stemmed from 1939 and was intended to be used in time of war or under comparable emergency situations. Putting the Rationing Law into effect was a first step toward the actual rationing of oil and oil products, but at the same time it provided the Minister of Economic Affairs with the legal basis for less radical measures, such as a ban on using cars on certain days. During the Suez crisis of 1956, the Drees Cabinet had also introduced car-free Sundays under the Rationing Law.

After consulting Den Uyl, Lubbers took up the mandate the Council of Ministers had given him the previous week. On October 30 he addressed the Dutch Second Chamber by means of a letter, in which he announced that November 4 would be the first car-free Sunday. In addition, he made various other suggestions for reducing petrol consumption, viz. through a voluntary speed limit of 100 kilometres per hour (ca. 65 mph) and by limiting the use of cars generally as far as possible. It was further important to reduce the use of heating fuels, by reducing the room temperature in homes and buildings and by closing curtains. Provided they yielded the desired results, Lubbers informed his audience, these mainly voluntary measures should obviate the need for stricter controls on distribution. Notwithstanding, the Cabinet had already prepared such rationing plans.[28]

On October 31, a debate was held in the Dutch Second Chamber in which both the international politics and the domestic aspects of the oil crisis were raised. Six members of the Cabinet were present at this debate: Prime Minister Den Uyl, the Ministers Lubbers, Van Agt, Van der Stoel, and the Under-Secretaries M.H.M. van Hulten of Transport and Water Management and A.P.J.M.M. van der Stee of Finance. The Cabinet took a cautious approach. Den Uyl outlined the government measures to the Chamber, characterizing them as 'precautionary measures'. Although up till then there had been 'not a single ton less oil arriving in Rotterdam', Den Uyl nonetheless suggested that the Arab embargo decision had been implemented, even though there was at that time no concrete evidence that this was so. Den Uyl was effectively bluffing, as one anonymous reader of the Foreign Ministry report on the crisis (cited earlier) later commented.[29]

Van der Stoel announced that no effort had been spared to get international consultations started to find an international solution to international problems, pointing to the OECD Oil Commission and the EC consultation process. 'It would be a very bad outlook for European business if such a common policy proved beyond reach.' For this reason, the Cabinet had proposed to its partners that they should consider the situation that had arisen at the coming EC sessions of November 5 and 6 which, as we already know, sounded much more auspicious than in fact it was.[30]

Lubbers announced that his policy should not only be aimed at maintaining the national provision of oil, but also the preservation of the international function of the Dutch economy, especially Rotterdam, the Rijnmond area and Amsterdam. He recognized that discussion within the framework of the OECD had not been very fruitful. Every country would first of all have to cut back its own use: 'only in a more drastically serious situation of real shortages would an international rationing scheme be put into effect'. The Netherlands therefore was on its own. The policy of the Cabinet was, he emphasized, primarily aimed at a 10% reduction of the national oil consumption.

It was the KVP parliament member M.W.J.M Peijnenburg who again accused the Cabinet of a rashness in its approach, while the great majority of the Chamber were in sympathy with the steps taken. There was only incidental dissatisfaction, such as over the fact that mopeds did not fall under the Sunday ban. Under-Secretary Van Hulten explained that a ban on mopeds would cause an enormous fuss and have very little effect. The government, however, stuck to its guns when the SGP member C.N. van Dis vainly asked whether church incumbents might not be included under a more flexible system of exemptions. The only suggestion that was immediately adopted from the Chamber was an amendment to the hours of the ban during the night of Saturday/Sunday: instead of the ban coming into force at midnight, it was delayed until 3 AM on Sunday. The Second Chamber then approved the proposals for economies in fuel use by a great majority, even though there were still no concrete indications of a reduction in the oil supply.[31]

The Rationing Law formally came into effect on October 31, the same day that announcements appeared in the *Staatscourant* – the official government gazette – informing the population that oil and oil products would be considered rationed goods and announcing the driving ban for

Sunday, November 4. The following day, at the request of the Second Chamber, that announcement was supplemented by the further decision that the ban should only take effect at 3 AM.[32]

During the Council of Ministers of November 2, it became apparent that certain ministers felt that Lubbers and his Ministry were pushing things too hard. Boersma said he was surprised at the speed with which measures had been taken; but Den Uyl parried this criticism with the observation that the Council had for this reason empowered Lubbers the previous week to prepare the regulations. Lubbers had rung him during the weekend to say that policy had to switch to action. Intensive discussions at the Ministry of Economics over the Sunday and Monday led to the conclusion that a rapid introduction of rationing was inevitable.[33]

After further thorough discussion of various other measures, it was decided to set up an interdepartmental committee in order to look into the various aspects more closely. On this basis, the *Co-ordination Group for Oil Crisis Action* was instituted several days later. The Ministers also decided that there should be preliminary talks between Home Affairs, Justice, and Transport and Water Management if a legal maximum speed limit were going to be introduced. In view of the possibly drastic consequences of the crisis for the national economy, the question was also raised of the relationship between government and trade unions as social partners. The Dutch Trades Union chairman Wim Kok had already informed the Cabinet that the Social Economic Council would have to consider how they should respond to the consequences of the oil crisis and that this would have to be further discussed with the government. If it should turn out that the economic foundations of the Cabinet's policy were affected, this could reasonably be discussed with the trade union movement. But the crisis need not necessarily interfere with the conclusion of a general agreement.[34]

Yet critics of the haste with which the Rationing Law was implemented were still not satisfied. The law itself was thought by some to be an antiquated and cumbersome instrument for dealing with the problems that had arisen. Van Agt sent Lubbers a letter in which he wondered whether rationing was a satisfactory instrument to control restrictions on the use of oil products.

> The bare fact that this law originated 35 years ago and was in principle intended for circumstances very different from those of the present may give rise to doubt.

An important reason for exercising caution was, for him, that the Rationing Law seemed a less suitable basis for taking criminal action against infringements of the Sunday ban on motoring. Van Agt therefore would have preferred creating a one-off, short-term law for rationing and restricting the use of oil products.[35]

The First Car-Free Sunday

Compliance with the motoring ban on November 4 was on the whole good, probably because restrictions on use of fuel were accepted as necessary by the general public and because similar measures had been announced in other countries. And furthermore, it was reasonable. The contrast between the silence and the usual commotion on the roads was much more impressive than in 1956. There also noticeably arose a certain national solidarity, partly because even Queen Juliana was publicly making use of the bicycle. In general, this first car-free Sunday was reported by the press in a lighthearted, cheerful vein.

Practically the whole millions-strong army of motorists had observed the Sunday ban in exemplary fashion, reported *De Telegraaf*. The Netherlands railway journal, *De Nederlandse Spoorwegen*, remarked that the passenger total was 30% up on a normal Sunday. Church attendance scarcely suffered at all: a poll in fifty different places showed only the slightest fall in the number of worshippers. 'In front of every church stood masses of bicycles and mopeds', according to *De Telegraaf*. The border posts had no exaggerated influx of foreigners to process.

> Quite a few Germans were rather surprised at the empty roads and took .... no notice of the recommended maximum speed of 100 km per hour. They were unaware that practically all petrol stations were closed.[36]

In Amstelveen, some forty boys and girls armed with sleeping bags, blankets and musical instruments took themselves to Motorway 6 to hold a picnic. After half an hour 'the youngsters were driven off the highway by the police'.[37]

'Sunday without deaths on the road' ran the headline of Monday's *Haagsche Courant*. And indeed this was one of the remarkable aspects of the Sunday. Four passengers seated in a carriage were reported injured when their vehicle overturned after the horse bolted on the Brienenoord Bridge in Rotterdam. The *Haagsche Courant* was also annoyed by those foreign motorists (to whom the Sunday ban did not apply), remarking that

> the Dutch highways were made rather unsafe by foreign – more specifically German – drivers who abused the freedom of the empty roads to drive at great speed; endangering all those children who had taken over the asphalt on their roller skates.[38]

But apart from this, it had been fun. Not that the press had no criticism to voice; but this disapproval was more of a continuation of the view cited earlier that the government measures were too limited. *Het Parool* argued for oil and petrol rationing, since the car-free Sundays affected certain economic sectors disproportionately. Rationing gave the public the opportunity to choose for themselves when they wished to use their cars.[39] And in the columns of NRC *Handelsblad*, the government was once again attacked for continuing to underestimate the gravity of the situation.[40]

However simple the action appeared at first sight, the first car-free Sunday did demonstrate how much administrative work was needed to restrict the use of oil and oil products. Not everyone turned out to be prepared to contribute as a matter of course; there were in all 120,000 requests for exemption. In fact, 15,000 exemptions were granted for motor vehicles. W.Q.J. Willemsen, who played an important official role in limiting oil use, can recall 'the most grotesque reasons and shameless arguments produced' to justify some of these requests for exemption.[41] The sheer quantity of requests and consequent orders gave an indication of the administrative rigmarole that would result if rationing were introduced. With subsequent car-free Sundays the number of requests for exemption only increased until, in the end, in the weeks leading up to the beginning of January, 74,000 exemptions had been granted.

During a discussion between the Attorneys-General, however, it became evident that the first car-free Sunday had not given rise to many problems, with relatively few infringements and impounded vehicles. In his first report, Rutten, chair of the Co-ordinating Group for Oil Crisis Action, reported that there had been 301 summons issued, and 239 vehicles impounded.[42]

It also turned out that roughly 90% of motorists were sticking voluntarily to the 100 kms per hour speed limit, encouraging the Attorneys-General to speak out against the introduction of a compulsory speed limit.[43] This latter judgement, in fact, was adopted by the Ministry of Justice, largely on the basis of the argument that the voluntary limit had yielded excellent results, whereas a compulsory speed limit would cause too heavy a burden on both police and the courts.[44]

According to the Attorneys-General, the biggest problems that occurred on the first car-free Sunday were related to traffic entering The Netherlands from abroad. Because some border posts were in fact unmanned, Dutch motorists entering the country were unable to get the necessary stamp that would allow them legitimately to proceed home via the shortest route. Another problem arose from the fact that in Limburg there were quite a few Dutch driving around in the cars of their German employers; and Dutch military personnel returning from Germany similarly caused problems because some Group Commanders had issued exemptions of questionable validity.[45]

Shell Helps

With the implementation of the embargo, and particularly with the participation in the embargo of Saudi Arabia on November 2 plus the restrictions on production announced two days later, the Dutch government – and specifically Economic Affairs – began to prepare for the worst possible scenario. At a staff discussion meeting on November 5, it was decided, in consultation with the Central Planning Bureau, to look at what might be the consequences of a 25% reduction in oil use, and what measures would be necessary in such a situation. What would happen if the oil supply to Rotterdam and the Rijnmond were reduced by 50% also had to be calculated.[46]

The situation that had arisen demanded especially close cooperation with the oil companies, however distasteful this proved to some backbenchers of the progressive parties. As Den Uyl told a meeting of the PvdA party executive committee:

> it was unavoidable that the Cabinet should have to form a common front with the major oil companies because as far as oil provision was concerned, the country was dependent on the oil companies.

Den Uyl later revealed that the government of course was concerned with procuring a fair, proportional allocation, but at the same time 'the suspicion that the oil companies might have a positive interest in an oil scarcity and its concomitant price rises'.[47]

Throughout the whole oil crisis consultations were held in various places between the oil companies and the Dutch government; most intensively with the Directorate-General for Energy at Economic Affairs, but regularly too with the Foreign Ministry. In any case, there was a school of thought, certainly within the Foreign Ministry, that this contact was not always properly co-ordinated and sometimes led to different conclusions in different places.

In the first place, these talks mainly boiled down to the fact that Shell and other companies 'were sounding the alarm vigorously' to get the government to act more positively. Indeed, it was partly on the basis of information passed on by the oil companies that Lubbers switched to the introduction of rationing. [48] The judgement of Rosenthal and Scholten seems correct, since on November 1 Van der Stoel informed the Dutch Ambassador in Brussels that the oil companies were suggesting that the situation was serious. [49] Other parties involved also believe that the companies pressed for use-restrictive steps to be taken urgently.[50]

It was clear by the beginning of November that the Dutch Cabinet was to a large extent dependent on the information provided by the oil companies. Moreover, it was in fact dependent on whatever policy the major oil companies pursued in the sharing out of the anticipated shortages. Initially, this question was regarded with the usual scepticism within the Cabinet, particularly by Van der Stoel, but on November 4 Wagner publicly gave an assurance that all consumer countries would in principle receive the same percentage less oil. This would have to be achieved by importing extra oil from those countries that had not joined the embargo nor imposed restrictions on production.[51]

Some of the major oil companies had besides already made preparations for limited production even before the oil crisis broke. Shell had presented most government leaders with a confidential 'Pink Book' identifying possible restrictions on the oil supply and outlining the chances of an 'oil scramble'. Shell was at that times of the opinion that in times of crisis stocks should be shared out fairly.

> Unlike the American companies, Shell had been campaigning for an intergovernmental agreement to share supplies in a crisis and, indeed, had already begun outlining, in its planning group, how such a system might work.[52]

With regard to the embargo against The Netherlands, the oil companies – or at least the European companies – thus in fact adopted a position of greater solidarity than The Hague's EC partners. It was Shell and BP that did not walk away from The Netherlands. The supply of crude oil by the American majors probably slumped more strongly than that from Shell and BP. In The Hague it was feared that the supply from Chevron and Texaco might be withdrawn entirely and that the refineries of these companies might even be shut down.[53] But Shell and BP, indeed from the very beginning of the crisis, proved themselves solid and reliable. The British government, in fact, put pressure especially on these companies to continue supplying as per contract – and by implication therefore leaving The Netherlands to sink. The same happened – according to Wagner – in Paris. It was therefore concluded at the Dutch Foreign Ministry, as a result of a conversation between Den Uyl and Wagner, that for the time being little could be expected from The Netherlands' EC and OECD partners. The Dutch stood alone, according to the Head of the Department for Economic Cooperation of the Foreign Ministry.[54]

Political-economic cooperation within the OECD or EC thus offered scarcely any point of contact for the Cabinet, whereas the attitude of the multinational oil concerns, so criticized in *Turning Point '72*, seemed to offer far better perspectives: they offered good cooperation. At Economic Affairs too it was quickly assumed that The Netherlands would benefit most from a course of events that left sharing the oil between consumer countries to the oil companies. The Department for Economic Cooperation at the Ministry of Foreign Affairs agreed, particularly in view of Shell's strong position in the international oil sector.[55]

The Hague thus had an interest in maintaining the free trade of the main oil companies – the majors – and of course the government also took risks, for it was by no means clear how these companies would react in the event of more serious shortages. The Department for Economic Cooperation of the Foreign Ministry therefore proposed that the government ought to be consulted more in the policy and planning of Shell. It had up till then not appeared a simple matter to get a clear picture of Shell's attitude, not least because Economic Affairs and Foreign Affairs appear to have had access to different sources which apparently did not always furnish the same information.[56]

November 7 brought new, reassuring, news over the policy of Shell and BP. During a meeting between Dutch Shell Director E.C. Werner and several top officials from the Dutch Foreign Ministry, it was made clear that both Shell and BP took the position, in defiance of the pressure from the British government and Prime Minister Heath personally, that the oil in Europe must be shared fairly. It was their intention to observe the measures imposed by the Arab countries, but nevertheless to achieve a redistribution by rescheduling to Rotterdam oil from Nigeria and Iran, which previously had gone to the UK. In spite of intense pressure from the new British Minister for Energy, Lord Carrington, Shell stuck to its principle of non-discrimination, a policy which caused great irritation in the UK. The London *Evening Standard* vented this displeasure with the headline: 'The Dutch are getting British oil'.[57]

Werner assured his Foreign Affairs interlocutors that Shell would not be pressurized by the British government. In the short term, however, it would be difficult to achieve an increase in production from Iran and Nigeria: in the preceding period, production in Nigeria had already been tripled and in Iran doubled. Shell had in fact decided to experiment with the transit of oil from one of the embargo countries via Rotterdam to Germany. If this worked out, Rotterdam could be rather satisfactorily organized.[58]

This *rescheduling* or redistribution of the oil flow entering Western Europe was to be an important means of neutralizing the effects of the embargo in the following weeks. Of course, this rescheduling had to be done discretely. From the beginning of November, therefore, there was far more reticence in the public mention of any figures relating to the oil supply. This did not facilitate the task of keeping the public informed, but it was nonetheless considered necessary, with a view to possible problems that might arise from the rescheduling of oil. Facts and figures concerning the oil supply had an enormous political significance.

In this context, it should also be pointed out that some Arab countries apparently had no interest in checking to see that oil tankers were not in fact supplying embargoed countries. Uncertainty over the source of oil also enabled the embargo to be circumvented. Subsequently, Wagner was to say that Shell had throughout observed the oil embargo and that The Netherlands was provided by countries that had not announced an embargo, [59] but it certainly remains an open question whether 'black' oil was also arriving in Rotterdam during the embargo. After all, that also happened in the USA, which managed to obtain leaked Libyan and Saudi oil during the embargo.[60]

Whatever the truth of the matter, the Head of the Department for Economic Cooperation (DES) at the Foreign Ministry observed that for obvious reasons Lubbers could only give the press general figures for the dependence on Middle Eastern oil; and in any case the population was sufficiently convinced of the seriousness of the situation. In confidential talks with representatives of the EC countries, figures for the inflow and transit of oil via Rotterdam had to be given, albeit general figures in the main, in order to demonstrate to the surrounding countries how important this supply and transit were for them. This was necessary to induce a certain degree of solidarity in these countries. Data on the division of the Arab oil supply between the European countries was highly confidential. According to DES, Shell had sought

> not to make these figures too clear, since they can then be changed in due course in favour of supplies from countries not involved in the embargo.[61]

Although there had initially been a feeling mainly of outrage in The Hague over London's attitude, it has to be conceded that all the West European countries switched to the defence of their own interests. This also caused conflict within the Benelux, a conflict in which the difference of outlook between the Dutch ministries of Economic and Foreign Affairs was again revealed. As we saw earlier, Economic Affairs wanted oil exports, with particular application to the considerable export to Belgium, to be subjected to licenses. So far this had not happened, because of the provisions of the Benelux treaty, although Brussels had in the meantime taken precisely such action. Pressure for discussion with the Belgian Ministry for Economic Affairs had been in vain, so now the Dutch Ministry for Economic Affairs also wanted licences as a counter-measure: export to Belgium by the oil companies must be limited, furthermore, and thus brought into line with domestic restrictions.

The Dutch Foreign Ministry, however, was not convinced of the wisdom of this kind of counter-measure. It seemed to the Head of DES that a better approach would be for the Foreign Ministry, via the Dutch Ambassador in Brussels, to convey to the Belgians their disappointment with the way things had gone. He also pointed to the risky aspects of the capped export restriction that Economic Affairs advocated, given the Dutch efforts to maintain a common market. This, after all, was the very thing The Netherlands reproached the British government with: exerting pressure on the oil companies. For these reasons, the Dutch Cabinet carefully measured any actions with regard to Shell. Foreign Affairs was right: restricting transit was a very delicate question. In the following weeks, this would seldom be explicitly referred to or discussed, not even in confidential interdepartmental talks. Economic Affairs no longer raised the question because, it was assumed at Foreign Affairs, nothing could be said about this to the outside world.[62]

The Co-ordination Group

Slowly, The Netherlands began to get to grips with the logistics of the oil crisis. The Rationing Law was meanwhile in operation, and in different places official consultative bodies came into being to prepare and execute the government's actions. In early November, in accordance with an earlier decision by the Council of Ministers, the interdepartmental *Co-ordination Group for Oil Crisis Actions* was created with Rutten as chairman. Meanwhile, in the Ministry of Economic Affairs the *Oil Crisis Working Group* was established as departmental co-ordinating body for the preparation and taking of all measures within Economic Affairs; beside which this body also served as advice group for the Economic Affairs representative in the Co-ordination Group mentioned above. The Directorate-General for Energy of Economic Affairs also served this particular task to a significant extent.[63] The Council of Ministers also prepared itself for this same function by creating a *Ministerial Committee for Oil Problems*, whose task was the preparation for decision-making within the Cabinet.[64]

November 8 saw the first gathering of the Co-ordination Group. Meanwhile, the reduction of oil production carried through by the Arab OPEC countries had risen to 25% compared to September levels; yet for all that, it was also acknowledged that the consumer countries had achieved no unity, partly in view of the privileged position of countries such as France and Great Britain. The OECD and the EC for the time being limited themselves to stock-taking and consultation.

At this meeting of the Co-ordination Group, the following principles were established for any further measures that might need to be taken. First, priority to be given to economic activity above private transport, and concomitant preference for the industrial fuel naphtha over petrol for motors. With regard to business life, a general restriction on use was thought best combined with a system of exceptions. This ought to lead to a global reduction in deliveries of approximately 15%. In the first stage, priority would also be extended to horticulture and public transport.

The Co-ordination Group argued for maintaining the Sunday ban on driving, even though it was clear that at Economic Affairs, and especially within the Directorate-General for Energy (DGE), there was strong support for the more radical measure of rationing. A driving ban on Saturdays, however, was thought by the Co-ordination Group to be too damaging to the catering industry. For the time being a strict policy on exemptions would have to be enforced that would make allowance for doctors, invalids and public transport. On the first Sunday, the ban had not applied to foreigners, but from November 11 they too would be included. Beside these regulations, the appeal to the people must continue for a voluntary speed limit of 100 kms per hour, to use their cars as little as possible and to restrict the use of heating fuel.[65]

Conclusion

Thus, the Cabinet prepared itself for the worst. There was great and continuing uncertainty over the actual effect of the Arab actions. Only by the third week of November would it actually be possible to say whether there was any reduction in supplies. Concerns gradually increased, however, that by the end of November would reach serious levels. But Shell, at least, had in the meantime declared its willingness to lend The Netherlands a helping hand. Although it could not be taken for granted that this pledge would be fulfilled under all circumstances, as far as the embargo against The Netherlands was concerned, the oil companies seemed to adopt a stance of solidarity that was all the more striking for its absence among EC partners.

Under these unclear circumstances Lubbers and Economic Affairs decided to opt for measures that would restrict oil use. Despite all the criticism in the press at the time, in retrospect the Cabinet can therefore hardly be accused of being indecisive. There was, after all, still no certainty that the embargo against The Netherlands would be effective. Den Uyl's suggestion in the Dutch Second Chamber that the Arab embargo was already being implemented was premature and in fact a bluff. It was mainly thanks to the influence of the Directorate-General for Energy and the oil companies that the restrictions on use were implemented so quickly. At this stage, the government's measures were without doubt considered by the public as legitimate. Apart from the many attempts to gain exemption, there was a high degree of readiness to cooperate. The appeal for voluntary speed restriction was also given public assent on a wide scale. People and government were in accord, although this would be a rather different story later when it came to the introduction of rationing.

The relative success of these counter-measures, however, could not obscure the fact that the oil restrictions hurt. On November 6, 1973, E.F. Geessink, the Director-General for Agriculture and Food Supply, of the Ministry for Agriculture and Fisheries, wrote in several letters to Economic Affairs that alarming news had reached him to the effect that supplies of petroleum to market gardeners were drying up. This meant that thousands of market gardeners were facing serious disruption to their cultivation schedules and the threat of losing their crops. Serious difficulties were also threatening the fishing industry, with consequent dwindling of the fish supply.[66]

5
A European Summit

The political divisions within the EC were further accentuated during November and December. France tried to exploit the crisis to press through a common EC energy policy and, moreover, with the support of London, to pursue a European-Arab dialogue. Washington also began to get more directly involved in managing the oil crisis. The differences between the Nine reached a climax during the EC Summit held in Copenhagen on December 14 and 15. Although there were also hopeful reports reaching The Hague, the situation for the Dutch during the weeks from November 6 to December 14 seemed worse than it had been throughout the whole crisis. It was in fact during this stage, as we shall see in Chapter 6, that tough domestic measures were decided, including the issue of rationing coupons. In this chapter, we shall look at the developments from the EC meeting of Foreign Ministers on November 5 and 6 up to the European Summit in December.

The Embargo

As we have seen earlier, with Van der Stoel's signature on the EC declaration of November 6, the strongly pro-Israeli character of earlier Dutch policy had been rather watered down. It was initially assumed that signing this conscious declaration would make a favourable impression in the Arab capitals; but the report of this declaration was swiftly followed by rumours of the interpretative statement that Van de Stoel had put out the same day. The Dutch representatives in the Arab countries were instructed to emphasise the strong unity of the Nine. But despite this effort, the positive effect of the new EC declaration remained limited; nor did the position of the government and Van der Stoel, in some respects, become easier in their own country. They were now under fire from two sides: from those who thought that Van der Stoel's pro-Israeli gruff lack of subtlety had put the national economic interests in danger, and also from those, like the historian L. de Jong, who thought the Cabinet had laid Israel open to Arab oil blackmail. Both sides demanded further explanation from the Cabinet of the signing of the EC statement of November 6.

The Arab side too, on several occasions, asked for clarification of the Dutch standpoint as a condition for any end to the embargo. In Tripoli, for example, J.J. de Roos, the Dutch Ambassador, learned that as it stood, the position was far from adequate.[1] The Dutch Ambassador to Egypt, Von Oven, was in fact doubtful whether the Dutch position had been improved at all by the EC declaration. With the approval of The Hague, he had held talks with the Secretary-General of the Arab League, M. Riad, who had suggested that it would be useful if Van der Stoel were to send a letter to the League pointing out once again that The Hague's standpoint did not deviate from the November 6 declaration. Moreover, according to Von Oven, Van der Stoel could also pledge that The Netherlands would increase development aid to the Arab countries.[2] Three days later, Von Oven repeated his message that a separate statement should quell Arab doubts about Dutch sincerity,[3] a suggestion which evoked only a negative reaction from Van der Stoel. He declined to address himself to the Arab League separately, since he had no wish to undermine attempts to arrive at concerted EC action.[4]

Several days later, the International Federation of Arab Trade Unions demanded that The Netherlands should put out a separate Middle East declaration. This proposal was endorsed by Egyptian officials, claiming that The Hague had only acquiesced in the joint EC declaration under duress. From Cairo, Von Oven warned of a hardening of the Arab attitude. The Arabs saw the embargo as an effective weapon: any 'lifting or softening its conditions would demand an even higher price'. In any case, Van der Stoel rejected this new demand just as he had declined to entertain similar Saudi demands (see the following section). The declaration of November 6 contained a common standpoint, according to Van der Stoel, and EC solidarity would only be damaged if member states began issuing their own separate statements. Syria and Jordan were also adopting a more anti-Dutch tone by mid-November, not least because of repeated expressions of sympathy for The Hague from Radio Israel. There were also various rumours circulating that served to stress the anti-Arab disposition of The Netherlands. For example, according to *Le Monde*, Van der Stoel had originally shown the door to the Arab Ambassadors on October

17. When two members of the Dutch Parliament questioned the Minister on this, he retorted that the report was a malicious distortion of the facts.[5]

But there were also more favourable signs. On October 17 the OAPEC countries decided in Vienna not to go ahead with the 5% reduction of oil deliveries to Western countries projected for December. The decision was considered by various commentators as an Arab token of recognition of the November 6 declaration. At the same time, however, it was decided to continue the embargo against the USA, The Netherlands and Denmark, and a week later, at an Arab summit in Algiers, to add to the list Portugal, South Africa and Rhodesia because of the colonial or apartheid politics of those countries.

The question, however, was whether the embargo against The Netherlands was effective. The Dutch Ambassador Schorer reported from Kuwait that the Kuwaiti Oil Minister Atiki had suggested at a press conference that Kuwait was not much concerned about attempts to circumvent the embargo. Kuwait, Atiki said, was in no position to control the exact destination of oil exports, or whether the destination of tankers was changed once offshore. So long as such manoeuvres were discretely carried out – 'without irrritating anyone' –, Kuwait had no objection.[6] Several days later, De Ranitz confirmed from Paris that both Kuwaiti and Saudi Ambassadors had let the Quay d'Orsay know that

> as far as they were concerned, oil originating from their countries and reaching countries affected by the embargo could be considered beyond their jurisdiction.[7]

Similar reports were coming out of Oman. It was a matter of assumption in Oman that The Netherlands would be able to look after itself by means of *rescheduling*. The same kind of reassuring communication was also coming from Bahrein;[8] all of which confirmed the conclusions drawn by De Lavalette as a result of his tour of the Arab capitals.

The measures taken by the OAPEC countries thus appeared much more stringent than they actually were in practice; and this was true not only of the embargo but also the restrictions on production. Thus, the news from Bonn was that, according to informants from the steel concern Thyssen, there was no less oil being shipped out of the Persian Gulf than normal. Supplies to established clients were perhaps down by as much as 25%, but this 25% was now being supplied to hitherto unknown traders. This oil was being offered at fancy prices, but some West German companies, Thyssen included, were willing to comply.[9]

Various offers were also arriving in The Netherlands. Ambassador Von Oven in Cairo, for example, was approached by an Egyptian consultant who wanted to know whether The Netherlands might be interested in 250,000 tons of crude oil. According to the documents this oil should have been shipped to Romania, but in reality it could go directly to The Netherlands.[10] Even bigger offers were received: Ambassador Boon reported from Rome an offer of more than 80 million tons of oil, with Beirut as the place for concluding the contract.[11] The Ministry of Foreign Affairs archive contains a series of such propositions involving, in various cases, at first sight rather louche figures who materialised as intermediaries – arms dealers and the like. Prince Bernhard also reported connections who would be able to assist The Netherlands to locate extra fuel.[12] At the end of 1973, there even came an offer from Baghdad. The brother-in-law of Saddam Hussein, Khalid M. Saloom, offered to supply two to five million tons of crude oil from the Rumailah oil field in Southern Iraq. Circles round Saddam apparently thought there had been enough pressure on The Netherlands and that the moment had now come to relax it a bit.[13] In general, however, these offers were considered by the Directorate-General for Foreign Economic Relations in Economic Affairs and by Shell as thoroughly unreliable. Nothing was done in response – which in itself would suggest that the shortage of oil was certainly not acute.

But despite all the offers of oil, by the end of October the supply of oil to Rotterdam did begin to decline. The Dutch Ministry of Economic Affairs anticipated that this shortfall would persist through the coming weeks, and it was even assumed that it might reach as much as 40 to 50%. In retrospect, these figures seem highly exaggerated; but the mood in The Hague at the end of November was highly pessimistic. On December 1, a sombre Den Uyl addressed the Dutch people on television. The world had irrevocably changed, he told viewers: the familiar times of before the oil crisis would never return.[14]

The Van Roijen Mission

Meanwhile, the Dutch Cabinet had decided on November 2 to send a special mission – in a certain sense a 'royal' mission – to Saudi Arabia in the person of the ex-Foreign Minister and one-time Ambassador J.H. van Roijen. This mission, like that of De Lavalette, was given as little publicity as possible; nor was it discussed in advance with any of The Netherlands' EC partners. Van Roijen's task was to convert the improvement in the Dutch image, possibly the result of signing the EPC declaration, into more concrete advantage. On the face of it, this would seem to have been no easy task, for King Feisal had taken serious offence to The Hague's response to the ultimatum of October 26. The intention was for Van Roijen to carry with him a letter from Queen Juliana, personally addressed to King Feisal, and whose text was discussed at great length both at the Foreign Ministry and by Shell's Arabists.

On November 15, accompanied by the head of the African and Middle Eastern Department of the Foreign Ministry, Van Roijen paid a visit to King Feisal and one of his counsellors. Van Roijen tried to elicit from the King some understanding of the Dutch position, but that was not a simple matter. Feisal maintained that The Hague had first to condemn the Israeli aggression publicly before there could be talk of any end to the embargo. Van Roijen's reference to The Hague's signing of the November 6 declaration made no impression. Feisal pointed out that The Hague had behaved in an 'oppositional' fashion in the meeting of October 13. Van Roijen denied this: The Hague had merely made 'procedural' objections at the settling of the October 13 declaration, and apart from that, both the October 13 and November 6 declarations of the Nine had been wholly unanimous. Feisal was not satisfied with this and laid down the standpoints that The Hague had to endorse publicly: (1) the condemnation of Israeli aggression; (2) demand for complete Israeli withdrawal from the occupied territories; and (3) support for the legitimate rights of the Palestinian people. He gave Van Roijen a letter addressed to Queen Juliana in which these demands were once more reiterated.

Van Roijen also spoke again to Yamani, who adopted a less punitive attitude than his monarch. The Oil Minister emphasised that the 'Saudis are being dragged by, rather than leading, the decisions of Arab countries concerning oil as a weapon'. In his report, Van Roijen concluded that the visit was appreciated and might perhaps contribute to Saudi Arabia refraining from any further retaliatory measures. The fact that Van Roijen had been received by King Feisal could only be of positive benefit to The Netherlands. Van der Stoel felt the same about the trip and thanked Van Roijen for his efforts.[15]

But the problem of the embargo was of course no nearer resolution. It was at this stage in mid-November that Van der Stoel became wholly convinced that the embargo had already been prepared well before the October war and was aimed against the whole of Western Europe. The analysis of Ambassador Boon, Dutch Ambassador in Rome, seemed to confirm such suspicions. In his analysis, the embargo plan must have been worked out on July 6 in Algeria on the assumption that the most effective action against Western Europe would be first to paralyse the working capacity of Rotterdam. This strategy was supposed to have been worked out under the leadership of a certain Ali Khodja, representative in The Netherlands of an Algerian oil company, Sonatrach.[16] On the basis of this information, Van der Stoel felt his view confirmed that the embargo was therefore not, or at least was not primarily, a measure aimed against The Hague's Middle East politics. The core of the conflict concerned Rotterdam and thence the whole of Western Europe. The war had certainly begun unexpectedly, but not the embargo. 'The scenario had long been prepared,' according to the minister, and there was no point in pursuing any form of political *rapprochement* with the Arab countries.[17]

Some Dutch diplomats thought otherwise and advised a more conciliatory approach. Immediately after the Van Roijen mission, Ambassador Derksen sent an analysis of Dutch-Saudi relations from Jeddah, advising the Minister to follow up Feisal's questions and his criticisms. Of course, it was questionable whether giving a single guarantee to Saudi Arabia would mean that the embargo would be lifted, but he feared that withholding all response would only further prejudice mutual relations and expose economic interests in Saudi Arabia to uncertainty.[18] For obvious reasons, given his reading of the whole crisis, Van der Stoel was not susceptible to this kind of suggestion.

King Feisal's letter was not answered till February 1974, but an initial reaction to his demands was communicated by Van der Stoel to the Saudi Ambassador Nowilaty.[19] The Dutch Foreign Minister made it clear that he was unable to satisfy Feisal's first demand, condemnation of Israel, since The Netherlands was not party to the conflict and took a neutral position. The government had already spoken its position in the unanimous declaration of the Nine and in conformity with resolution 242. The Cabinet did not intend to reconsider this position. Van der Stoel again stressed to Nowilaty The Netherlands' determination never to condone acts of violence. The government subscribed to the view that the Palestinian question was a political matter, but it was not for The Hague to anticipate a future peace accord by specifying Palestinian rights.[20]

Van Roijen's journey did not therefore have much effect. The Dutch reputation seemed rather to have sunk to its nadir. Only Morocco, which in any case had no oil to export, seemed at this stage to adopt an attitude less unfavourably disposed toward The Netherlands. This was the message communicated by Ambassador De Vreede from Rabat following a conversation with the Moroccan Foreign Minister. According to the latter, the problem of how to get oil to The Netherlands discretely, for example through such intermediaries as France or West Germany, was now being studied in several Arab capitals.[21] This sounded encouraging, and the message was relayed by the Dutch Foreign Ministry to all diplomatic posts in the Arab countries. At the same time, however, it was still being insisted elsewhere in the Arab world, in Beirut for example, that the Dutch Cabinet must publicly condemn Israel before any action against them could be relieved. In a conversation with DGPA Van Lynden, the Lebanese Ambassador in The Hague again repeated explicitly that The Netherlands had to acknowledge two matters in a separate statement: 1) the rights of the Palestinians, and 2) the necessity for Israel to vacate *all* occupied territories.[22]

An Incident in The Hague

Meanwhile, in The Hague a second note was sent to the Second Chamber in which it was again stated that the government's policy with regard to the Middle East was even-handed. A solution to the Middle East conflict could only be achieved on the basis of resolution 242. The declaration of November 6 contained several points that elaborated on 242, in particular with regard to the political aspirations of the Palestinians. Now that there were genuine prospects for peace negotiations, 'the political aspirations of the Palestinians must be articulated in some shape or form'. The note proposed that the passage in the EPC statement of November 6 over the ending of territorial occupation was in complete harmony with the Cabinet's position. It was stressed that in the many contacts with the Arab countries, time and again there had been reference to the mistaken supposition that the Dutch had offered help to Israel in the form of arms deliveries or volunteers. The note repeated that the embargo had to be seen against the background of the changing economic relations within the oil sector. In other words, the embargo was therefore not, or at least not exclusively, a punitive measure directly related to The Hague's Middle East politics.[23]

At the end of November the policy was raised for detailed discussion during the general debate in the First Chamber. The signing of the EPC statement was in general endorsed, although ex-premier P. de Jong (KVP) thought it would have been more open and above board if a change in policy had been openly admitted. For the reality was that Van der Stoel had shifted ground for the sake of maintaining EC unity. In response to the views of the First Chamber, Den Uyl gave his minister his complete protection. The storm of criticism Cabinet policy was met with was in his view totally unjustified. The story in *Le Monde*, that Van der Stoel had more or less shown the four Arab Ambassadors the door, had no basis whatsoever. Den Uyl let it be known that one of the Arab Ambassadors concerned, on his own initiative and on behalf of his three colleagues, had approached Van der Stoel to tell him that he deeply regretted the piece in *Le Monde*. As far as the passage of the EPC statement was concerned that spoke of the legitimate rights of the Palestinians, Den Uyl acknowledged that the formulation here differed from that of resolution 242, but there had been no change of policy.[24]

The Foreign Affairs budget debates in the Second Chamber were scheduled for November 29. There, too, the accusation was heard that the change of policy implied by undersigning the EPC statement had simply not been acknowledged. In reply to questions about the rights of the Palestinians, Van der Stoel answered that their political aspirations had to be given form, a position which had meanwhile been endorsed by The Netherlands in the UN. It was self-evident, said Van der Stoel, that the Palestinians had to have some say in the realisation of their own future. How that should be done, however, was not in the Cabinet's power to decide.[25]

In the Parliamentary debate, Van der Stoel avoided the issue of handing back the occupied territories. Following the advice of DGPA Van Lynden, he referred to the relevant passage in the EC declaration.[26] This manoeuvring was not easy for Van der Stoel. It was being solidly maintained in the Second Chamber that there had been no change of position while, at the same time, in the higher realm of international diplomacy it had to be implied that by signing the EC declaration The Hague had indeed accepted the position demanded by the Arab countries, without this being spoken in so many words. This of course was the root of the problem: that the normally implicitly accepted diplomatic positions were not always backed by domestic or parliamentary consensus. This, in turn, led to complications like the so-called Thurkow affair.

The main source of the problems that arose in The Hague in early December lay in the cryptic passage in the November declaration concerning the occupied territories. The Foreign Ministry spokesman, Chr. Th. F. Thurkow, told the press that the EPC standpoint implied that Israel had to withdraw from *all* the occupied territories, with the qualification of possible minor border corrections. This statement was apparently made without prior consultation with his minister. Thurkow's pronouncement meant 'a shift in the Dutch position ... which may well have suited the views of DGPA Van Lynden, but certainly not mine', Van der Stoel later remarked.[27] Thurkow's interpretation, which was moreover already being taken up in diplomatic negotiations, caused great commotion. On Christmas Day the Second Chamber called Van der Stoel to account.

During the debate, although Van de Stoel insisted that his policy had undergone no change, he did not explicitly distance himself from Thurkow's interpretation. He pledged that action would be taken to ensure that further misunderstandings would be avoided. In a television interview the same evening, he reaffirmed the remark of Den Uyl who, in an interview with the French newspaper *Le Monde*, had said that he thought the definite article '*the*', defining 'occupied territories', was not all that important. Assuming the possibility of minor border corrections, the discussion over the English and French texts of 242 was in fact utterly pointless.[28] Van der Stoel seemed to be partly covering for Thurkow, but nevertheless his spokesman was suspended the following day. There was no doubt in the press, however, and with good reason, that the whole Thurkow affair was the result of the government and Van der Stoel continuing to deny that the Dutch signature to the EC declaration involved a shift of policy.[29]

The day following the debate in the Second Chamber, opposition leaders H. Wiegel (VVD) and R. Kruisinga (CHU) had a meeting with Den Uyl and Van der Stoel, with opposition spokesmen for foreign affairs, H.J. Koster and D.F. van der Mei, also present. The two opposition parties thought the Cabinet should be much more explicit in acknowledging that the EPC statement of November 6 meant a change of policy. They expressed the hope that the coming Summit in Copenhagen would provide the opportunity to clarify the European position. According to Wiegel and Kruisinga, the Cabinet's lack of clarity had given rise to confusion, while the EPC statement was clearly a revision of resolution 242 in a pro-Arab sense. They asked:

> Does the government not understand that its every pronouncement is front page news in the Middle East, that our Ambassadors Bentinck in Algiers and Von Oven in Cairo – to name but two – are deeply unhappy because they increasingly find their task becoming untenable?[30]

Both Van der Stoel and Den Uyl rejected this criticism. The Netherlands, said Den Uyl, had conducted a pro-Israeli policy for 25 years, and this was not about to change. Van der Stoel declared that he would oppose any clarification of the November 6 statement at the coming Summit. This of course did nothing to remove the impression that there was indeed a deliberate vagueness, or room for manoeuvre, in the Cabinet's attitude.

The European Community

At this stage, London and Paris remained resolute in their opposition to any activity aimed at sharing the brunt of the oil shortage on equal terms. In mid-November, mutual relations between EC countries appeared to become more acerbic than ever, with the Dutch Cabinet, and particularly Lubbers, contributing to this frosty climate. In the meantime, Lubbers had declared in the Second Chamber that he would not shrink from further steps, if necessary, to safeguard the national provision of oil. This could affect the export of natural gas, both the quantity exported and the price. There was no obligation on The Netherlands to be 'holier than the Pope' compared with other EC states. Nor did he exclude the possibility, he said, that The Netherlands might block decision-making in other areas within the EC. He did not intend this as a 'threat'; rather, in the absence of unanimity in this question of oil supply, it was a perfectly 'normal and level-headed observation that it would be senseless to pretend one could engage in actual, genuine consultations on other matters'. The PvdA Second Chamber member Van der Hek responded positively to this statement. There was also support for the minister in the press.[31]

The oil crisis opened up fundamental differences of outlook among member states over the future of the EC. True to tradition, it was above all the Dutch and the French conceptions that were at variance, as was evident when Van der Stoel paid a (long planned) visit to Paris on November 8. During a *tête-à-tête* with Pompidou and Jobert, among other matters the future of European integration was discussed. Both French politicians made it clear that they wanted a European summit conference, in any case before the end of the year, to discuss the question of whether or not the Nine really were serious about working for European unity. Van der Stoel assured his French colleagues once again that The Hague wanted European unity, but stressed that during this crisis it was also a matter of defending what had already been achieved; for the embargo threatened that achievement with disintegration. Pompidou – more than Jobert – appeared to endorse this view. As far as the European summit was concerned, a summit much desired by the French, Jobert emphasised that this would have to give directives to a future Energy Council which were totally in conflict with the usual Dutch position on institutional relations within the EC. Much to Van der Stoel's relief, Pompidou seemed to entertain less radical ideas. In the eyes of the President, it should be more of a *'tour d'horizon'*.[32]

The points of view put to Van der Stoel at this meeting by Pompidou thus seemed less objectionable than those put by Jobert, which in fact corresponded to earlier (and subsequent) impressions that it was primarily Jobert who seemed to advocate a more extreme viewpoint. Jobert was not popular in The Netherlands in those days. Ed van Thijn noted in his diary on November 7: *'Jobert, rien, rien, rien'*.[33] Whatever the case, it was clear that little sympathy could be expected from the Parisian side for the Dutch wish for joint EC action. During the course of November, moreover, rumours began circulating to the effect that Paris intended to force the oil companies by legal means to fulfil their obligations. At this stage, evidently, Paris had no intention of introducing any measures to limit consumption.[34]

Little support was expected from the British side too, as Den Uyl remarked a few days later. The situation in Great Britain was additionally difficult because of serious industrial unrest among the miners, which on November 14 led the British government to declare a state of emergency. As reported earlier, this was the context in which London was considering using legal measures to force the oil companies to supply Great Britain with its oil quota in full. The British Cabinet specifically put pressure on Shell and BP not to redirect to The Netherlands the Iranian and Nigerian oil that was on its way to England. In mid-November, when it was found that these companies were sharing out the oil shortage pro rata, the British press mounted a fierce attack on the oil companies who, according to *The Times*, were busy 'diverting some supplies to other customers'. The companies concerned were Shell, BP and Gulf.[35]

It was during a meeting of the Socialist International, for which Den Uyl had travelled to London, as mentioned in Chapter 1, that Golda Meir so warmly thanked the Dutch Premier for his government's support. The meeting was marked by a dramatic confrontation between Meir and the majority of West Europe's social democratic leaders, only Brandt and Den Uyl being excepted from Israeli reproaches. These were also the two countries that had actually supported Israel during the first weeks of the war. This aside, it emerged that the British Labour leader, Harold Wilson, objected to the November 6 EPC statement, indicating what might be the policy of a Labour government, which was in fact voted into office the following March.

At the time, however, the Conservative Prime Minister Heath occupied Downing Street. On November 11, Den Uyl held a memorable meeting with Heath, at which the two leaders disagreed about almost everything. Heath said the only solution to the oil problem was to persuade the Arabs through diplomatic contacts to give up their restrictions on production. Calling the Arab demands unacceptable was senseless. Nor was there any point in creating the impression that Europe was hostile to the Arab world. It was, in fact, essential that the Arab countries should have friends outside the Soviet camp. Heath declared his satisfaction with the November 6 EPC statement, specifically because of its direct reference to resolution 242. It was Israel's desire not to implement resolution 242 which was, after all, the root cause of the war. He appreciated that The Hague had signed the declaration and expressed his view that the EC declaration should now be followed by a suitable diplomatic mission. Den Uyl found himself entirely at odds with Heath's pronounced views. A common EC front would make it clear that the European countries were not prepared to bend to Arab blackmail. He stressed that the embargo against The Netherlands was in fact aimed at the whole of Western Europe.

It was clear that the two premiers were starkly opposed to each other. They similarly held different views on the question of how effective a grip the Arab countries had on compliance with the embargo, which was of course an extremely important question in deciding what strategy to follow. Den Uyl, according to his own report of events, pointed out that

> it is highly significant that the oil companies have and continue to have the freedom to regulate the supply of oil to different countries.

Heath, who may well have been astonished to hear this statement of principle from the social democrat Den Uyl, gave as his opinion that the Arab countries had a fairly good grip on the movement of tankers, but he did not contradict Den Uyl's opinion on the oil companies' freedom. In fact, it seemed to confirm the judgement of the British Foreign Minister, Home, who three days previously had said: 'We will not interfere with the commercial policy of the oil companies'.[36]

Subsequently, in the Council of Ministers, Den Uyl declared that Heath had fully supported the Arab countries and that he found him 'cynical'. Den Uyl believed that Heath's attitude was largely dictated by the huge Arab investment in the British economy. Indeed, Heath had openly admitted his greatest fear: that the Arabs would pull out their Sterling credit, which would be a heavy blow to London as a financial centre.[37]

London and above all Paris were at this stage without any doubt the most important opponents of the Dutch, as was observed by Bonn with some concern; so much so that in mid-November it was being wondered in the West German *Auswärtige Amt* – the Foreign Ministry – whether The Hague should not seriously try to improve relations with France. The Dutch Ambassador De Beus reported from Bonn that it was being wondered there 'whether we might not do well to be more obliging to the French in other official bodies – the UN, for example'. The implicit reference here was to the question of a moratorium on nuclear tests, in which EC countries had up till then adopted a position in the UN opposing France.[38] This was a suggestion which, as will become apparent, did not misfire.

The differences of opinion within the EC gradually crystallised, threatening an impasse. This proved to be the case when the Permanent Representatives of the EC met for discussion of the oil situation on November 19. The familiar moves were rehearsed. The Belgian representative observed that sharing the oil within the common market at that moment depended in fact on the arbitrary decisions of the oil companies. His French colleague said that these companies had insufficient political power to ensure safe supplies of oil in the longer term. Furthermore, this arbitration worked more to the advantage of the USA than the European countries. The French therefore wanted joint state intervention in a European framework and not within the OECD. The British representative concluded, however, that the problem was essentially political and therefore was only susceptible to a political solution, which should in the first place be sought at the EPC meeting of November 20. The British thus supported the French only in working for a political rapprochement with the Arab countries, but not if it was a question of regulating the market or of state intervention.[39] There was therefore at this stage absolutely no advance in the question of mutually supportive action within the EC. The same was true of the OECD.[40]

The pressure of the oil-producing countries was increased further by another round of price hikes. On November 19 it was learned that discussions in Vienna between OPEC and the oil companies had failed. The OPEC members let it be known that henceforth they themselves would set prices. Three days later, Wagner sent Den Uyl a letter expressing his serious concerns over the anticipated price rises. He appealed to the consumer countries to restrict the race for oil through mutual consultation and to warn the OPEC countries of the possible collapse of the world economy. Van der Stoel immediately informed all his embassies that the Cabinet shared the deep unease like Shell.[41]

In the context of Europe, the EPC meeting of November 20 had meanwhile produced a modest breakthrough with regard to the embargo. The British delegation proposed that a common démarche be undertaken in all the Arab capitals. The plan met with scepticism from the French, but they pledged to support the proposal, at least if the EC member states withheld their support in the UN General Assembly for a resolution condemning French nuclear tests, a resolution to which The Netherlands had initially given its support. This was the compromise that De Beus had earlier mentioned in his report from Bonn.

Van der Stoel immediately promised that his delegation would abstain when it came to voting on the resolution. The delegation would explain the reason for its abstention subsequently by alleging that it had to do with imbalances in the text. Bonn also promised to abstain on the same grounds, and with these pledges Paris agreed to support the British plan. Van der Stoel was later to say in the Council of Ministers that his promise over the nuclear testing was bordering on the limits of the acceptable. Nevertheless, there was now for the first time a united front presented by the Nine with regard to the embargo. Den Uyl also referred to Van der Stoel's performance as 'balancing on a knife-edge'.[42]

For the purpose of the common EC démarche, a joint text was subsequently put together in which it was claimed that the Arab actions had affected the interests of the entire EC. The Arab countries had let it be known that the EC should play a role in procuring a just and lasting peace in the Middle East. The embargo threatened to divide the EC, however. As a result, there was a certain contradiction between the implications of the Arab measures and the unanimous EC attitude that they desired.[43]

But the démarche did not produce the desired effect. Executing such a joint move turned out to be more complicated than anticipated. In every Arab country, the most influential Ambassador was to be entrusted to carry out the démarche; but this was not always regarded with confidence. The Dutch Ambassador Van Hoeve pointed out from Damascus that the démarche there was executed by the French Ambassador; while it was very much open to question whether he really would be prepared to waive French interests for those of the other member states. The same was the case in Algiers: the French Ambassador was to execute the démarche.[44]

And indeed, as Dutch Ambassador De Ranitz reported from Paris, it was quickly apparent that the French had given their own interpretation to the joint démarche in some Arab capitals. Specifically, the French Ambassador in Jeddah had received instructions that did not conform with the EC arrangement. The Director for North Africa and the Levant at the

Quay d'Orsay positively denied this to De Ranitz, who had left it at that, since he was unable to say 'that access had been obtained to the relevant instructions, which furthermore would certainly not promote the further cooperation of the French Ambassador in Jeddah'.[45] The result of this joint action was, in fact, very little.

The end product of all Van der Stoel's diplomatic activities and the work of the Dutch Foreign Ministry were thus so far negligible. By the beginning of December, the only thing that could be positively said was that most EC countries meanwhile, albeit in a rather formal and not entirely convincing fashion, had made it clear to the Arab countries that they objected to the embargo against The Netherlands. This was at least a step, but not a great step forwards.

Two Oil Ministers in Europe

Shortly after the EPC declaration of November 6, it was learned that two Arab ministers, the Saudi Oil Minister Yamani and the Algerian Minister of Industry, Belaid Abdessalam, were intending to visit several West European countries. There was huge agitation surrounding this projected visit, both in The Hague and in other West European capitals. As soon as the trip was known, attempts were made to persuade the two ministers to pay a visit to The Netherlands. On November 21, DGPA Van Lynden emphasized to the Algerian Ambassador to The Hague that such a visit was thought 'particularly opportune'.[46] The following day, Van der Stoel sent a coded telegram to all posts repeating that the Cabinet was eager to receive the two ministers, although he considered the chance small that they would actually visit The Hague. Van der Stoel suggested that other EC governments would be able to see that the refusal of the two to visit The Hague would cause surprise, since The Netherlands had taken absolutely no counter measure against the embargo.[47]

While the two ministers had already begun their journey and had called on Paris, in The Hague differences of opinion were revealed over the question of whether or not an urgent attempt should be made to arrange a meeting. During a discussion meeting between the Dutch Foreign Ministry and the Ministry of Economic Affairs, Lubbers expressed interest in meeting the two – once again an indication of Lubbers's more flexible and pragmatic approach than Van der Stoel's. According to the latter, Lubbers and Economic Affairs were indeed more pro-Arab, but that was hardly surprising in view of their justifiable concern for the economy. Lubbers contested this. He said that as soon as the embargo had been announced, in a meeting with Den Uyl and Van der Stoel, he had argued 'without reservation, not to give in an inch politically', which somewhat surprised the premier.[48] Nonetheless, the Foreign Ministry had doubts about Lubbers's proposal, since such a meeting should in the first instance be about political demands.[49]

In any case, on November 26 in Paris, Yamani absolutely denied ever having received a Dutch invitation. Van Lynden later forcefully reminded the Saudi Ambassador Nowilaty that he had on two occasions made it quite clear that The Hague would highly appreciate a visit from the two ministers.[50] On the same day, the two ministers let it be known informally that they were ready to receive a Dutch delegation during their stay in Brussels. Van der Stoel was fully informed of this by his Belgian counterpart, Van Elslande. The Belgian minister thought The Hague ought to respond positively to the Arab invitation; he feared the possibility of Rotterdam being eliminated and the consequences that would have for the Belgian economy.[51]

The visit to Brussels for which Van Elslande was arguing was raised for extensive discussion at the Dutch Council of Ministers. Van der Stoel had little enthusiasm for a visit by himself and Lubbers to Brussels. In all probability the two Arab ministers had no mandate to negotiate, he thought, and were therefore only interested in gaining concessions. It seemed to him, at least, that it would be better if he did not go himself, thus preventing any meeting with the two ministers assuming a political nature. Furthermore, the two had not taken up earlier invitations, even denying there had been any, which was also unpromising. However, Van der Stoel had no serious objection to Lubbers going. It would probably not be a good idea to reject the invitation from the two ministers completely, the more so since that would also be likely to irritate the Belgians. Lubbers was furthermore in a better position than himself, for he could steer clear of the political aspect.

Boersma (Social Affairs) and Pronk (Development Cooperation) also had little appetite for a 'humiliating trip' to Brussels, but on the hand, it was felt the impression should be avoided that The Hague was unwilling to talk. Pronk further thought that any visit Lubbers made would be useful in the context of the government's image with the electorate. Lubbers himself declared that he was ready to meet the two Arab ministers. In his discussions, he would above all point out the repercussions the embargo had for the EC, because Europe was in fact an integrated entity. Den Uyl thought that undertaking this 'humiliating visit' was indeed a fine balancing act. In the publicity given to the visit, it should therefore be stressed that Lubbers would visit his opposite number, the Belgian Minister for Economics, and that coincidentally there would be discussions with the Arab ministers. This was approved by the ministerial council.[52]

The meetings of the two Arab ministers with members of different European governments did not go off with the same facility. In particular, even though Van Elslande had meanwhile given the Belgian parliament an interpretation of the EPC declaration of November 6 that was favourable to the Arabs, the Belgian-Arab meeting was rather unpleasant.[53] The two Arab ministers let it be known that they expected from Europe deeds rather than words that would force the Israelis to withdraw more quickly from the occupied territories. Van Elslande made it clear to his Arab guests that he found the discrimination against The Netherlands unjust, which scarcely improved the atmosphere of the meeting.

This conversation between Van Elslande and the two ministers thus appeared to deliver very little, apart from one matter which, for The Netherlands, was highly significant. Having been informed of the Belgian ownership of the Rotterdam-Antwerp oil pipeline, Yamani showed himself interested in the possibility of getting oil to Antwerp via this pipeline. Out of consideration he said, he would like to look into this, at least if it could be guaranteed that the oil was in fact pumped to Antwerp and did not end up on the Dutch market.[54]

The following day saw the meeting take place between Lubbers and the two ministers. Lubbers, without a single political concession on offer, was accompanied by DGPA Van Lynden, so that in fact the Foreign Ministry was directly involved.[55] The meeting proceeded more or less along predictable lines. When the ministers once again asked for a separate statement from The Netherlands, Lubbers replied that there was no reason to assume that The Hague was not foursquare behind the statement of November 6. According to Van Lynden, this position was also set out in the note of November 16, as well as in Van der Stoel's account to the Dutch Second Chamber on November 29. Both documents, in translation, were handed to the Arab ministers. Van Lynden stressed that the EPC statement, when it came to the question of the occupied territories, left nothing to be desired as far as its clarity was concerned. To ensure that there should be no possible misunderstanding, Lubbers added to this that The Netherlands was 'opposed to annexation'. Finally, Van Lynden declared on behalf of his minister, Van der Stoel, that the Palestinian question was not purely a humanitarian question; The Netherlands recognised the Palestinians' *legitimate rights*.[56]

Lubbers and Van Lynden therefore went some way to meet the two Arab ministers; for Van Lynden's clarification and Lubbers's formulation of the Dutch attitude to annexation both implied Israeli withdrawal from *all* occupied territories. The view of *Trouw*, that Lubbers's attitude in Brussels had been 'a model of firmness and consistency upon which more reformative brethren might reflect in admiration', would therefore seem to be a little exaggerated.[57] Whatever the case, it was still not enough. The Algerian minister Abdessalam in particular still desired a public declaration that Israel must withdraw from all the occupied territories. Just as Van der Stoel had said, the two ministers only wanted concessions and were neither prepared nor able to offer anything in turn.

American Support

Throughout a large part of the whole oil crisis, there was talk of the American willingness, if it came to the crunch, to provide The Netherlands with extra oil or with the necessary financial resources. This backing was first explicitly expressed on November 7, the day after the EPC gathering, when the American Defense Secretary, James Schlesinger, paid a visit to The Hague and during the course of a meeting with Van der Stoel offered assistance with oil supplies.[58] The offer was kept secret from the outside world, but almost at once the buzz of rumours could be heard. Two weeks later it was learned that

> 'the Americans let it be known, reliably and at the highest level, that if the worst came to the worst, proposals would be put to the US government to help The Netherlands to deal with the consequences of the embargo'.[59]

According to Van der Stoel, Kissinger had said the various possible ways of doing this were being studied. The American ambassadorial counsellor, Steve Blodgett, informed DGPA Van Lynden that assistance would be given only if it were requested by the Dutch Cabinet. Moreover, the decision would also depend on whether or not a European oil-sharing agreement had been reached. Van Lynden assured him that such a request would only be made in the extreme case. Notwithstanding these arrangements, the Dutch Foreign Ministry was busy drawing up a damage assessment; no simple matter as it turned out, because it still remained unclear just what the effects of the embargo on The Netherlands were.[60]

The American offer was made against a background of increasing American-European discord over the Middle East peace process. The gulf separating Israel and Egypt was still enormous. On November 22, the Israeli Foreign Minister, Abba Eban, had discussions in New York with Kissinger and with UN Secretary-General Waldheim over a peace conference. On the same day, Israel and Egypt appeared ready to make concessions. Three days later, Tel Aviv announced that it was in principle willing to participate in a peace conference. Egypt also agreed. The peace conference would take place in Geneva on December 18. On November 13, American Under-Secretary for Foreign Affairs Sisco had already informed Van der Stoel that Washington saw no role for the EC at the planned peace conference, since the content of the November 6 declaration was incompatible with EC participation as an impartial broker. On November 21, Kissinger said he could only interpret this declaration, given its content and timing, as giving way to Arab blackmail. Once again a serious difference of opinion seemed to have developed within the Atlantic Alliance.

The American offer of assistance met with very different responses in the Dutch Council of Ministers, and was debated on several occasions. Although the Cabinet certainly did not warm to it, Lubbers was all for a positive response. He wanted to begin secret negotiations with the Americans in early January, so that the whole affair could still be considered in the light of an anticipated European solidarity. He realised that this was 'a political act'. [61] At the end of November, Van der Stoel put the argument that, in connection with this support, The Netherlands should not, in fact, 'engage in any policy that would conflict with American policy'. Vredeling, however, thought it more sensible not to take up the offer in view of the fact that it would align The Netherlands even more firmly with the USA and lead to even deeper isolation within the EC. His judgement was shared by Van der Stoel, though the latter felt the offer should not immediately be rejected. But the priority was indeed to find a formula within the EC that could give definite form to mutual solidarity there.[62]

The offer certainly had its disadvantageous aspects. It also raised doubts within the American presidential advisory body, the National Security Council (NSC), as to the wisdom of such a step: so reported the Dutch Ambassador in Washington. A supportive action might cause tensions within the EC and give the impression that Kissinger was out to fish in troubled European waters. The question posed by NSC staff was: given this background, to what extent would The Hague appreciate such help? The consequences for the Dutch position in Europe were, after all, not easily calculated.[63]

Indeed, there were even more doubts within the Dutch Cabinet. In any case, thought Van der Stoel, the whole affair must be kept out of the public eye. That, it turned out, was not easy. Lubbers had already publicly said that the government was in agreement with this support proposal. Furthermore, it was meanwhile learned that the US Under-Secretary, W. H. Donaldson, would come to The Hague in December. On the other hand, the offer of support strengthened The Hague's hand within the EC. In fact, at the end of the Council of Ministers, Van der Stoel informed the press that they had learned with gratitude of the American offer and awaited Donaldson's arrival with much interest.

The side effects of this possible oil assistance were subjected to thorough study at the Dutch Foreign Ministry. American support would presumably mean the definitive end of solidarity within the EC. Accepting help, furthermore, would give the impression that The Hague would come out of the crisis rather well off. The EC partners were by now already keeping a sharp eye on The Netherlands, since although the Dutch had been harder hit by the embargo than other European countries, they were nonetheless in a position to escape much hardship because of their natural gas. The Dutch position could become uncomfortable, for The Netherlands would then be considered in EC and EPC contexts as an American protégé in matters of oil and energy. In Arab eyes, The Hague would simply be seen as an American vassal. Indeed, it was not inconceivable that the Arab countries would react by instituting further punitive measures against Dutch interests. Even if everything went well, The Netherlands would henceforth always be seen by the Arabs in a poor light, separate from the rest of Europe.

In short: to accept American aid would be to accept a signal change in the 'balanced' policy pursued so far. The priority must be therefore to urge solidarity between the Nine, and a tougher attitude if possible toward those countries that would threaten the economic life of the Dutch nation. Only when all chances of a united European front had been exhausted should The Netherlands turn to America for help. And in the meantime, of course, as Van der Stoel had recommended in the Council of Ministers, the offer should be kept alive.[64]

The Dutch-American rapprochement appeared to be strengthened further when on November 30 Van der Stoel received an invitation for a bilateral discussion with Kissinger on the occasion of the coming NATO Council of Ministers conference. R.B. van Lynden, the Dutch Ambassador in Washington, thought Van der Stoel should accept this invitation, especially since Jobert was opposing a proposal for a meeting on December 9 between Kissinger and the Nine. Of course, there were also objections attached to such an American-Dutch meeting. There was always the chance that Kissinger would insist on Van der Stoel choosing whether to follow his policy openly, which would only exacerbate the tensions within the Nine.[65]

Naturally, this rapprochement between the USA and The Netherlands aroused considerable interest abroad. The German embassy approached the Dutch Foreign Ministry with the question of whether there were any more details to be told of the offer of oil and the discussions that Donaldson was to hold.[66] The offer of support (and of course the embargo itself) seemed to make The Netherlands the fulcrum of all the diplomacy of the oil crisis. An advisor to the Secretary-General of the Foreign Ministry, H. Scheltema, had the impression that the USA, possibly during the upcoming NATO session, would try to build a greater Trans-Atlantic solidarity against the Arabs using The Netherlands as its instrument. It did seem, according to Scheltema, as if The Hague was the focus of all diplomacy.

> The Arabs hope to pressure us, and Europe with us, into supporting their cause, the Americans, by supporting us, are trying to move Europe toward a less pro-Arab position. In this situation, I cannot see clearly what we can do for the time being other than continue to play the European card.[67]

And according to De Ranitz, the French Director-General for Political Affairs thought that the US had ultimate tactical intentions in offering oil, and was trying to prise The Netherlands out of the EC.[68]

In this connection, it is striking that the offer of assistance was regarded with considerable scepticism within the top levels of Shell. A.C. Helfrich, Director of Shell-Netherlands, later pointed out that the American market was itself threatened with the possibility of major shortages, making it extremely unlikely that Washington would actually support The Netherlands in the case of a serious reduction in oil supply.[69]

Divisions within the EC

As we saw earlier, Paris still wanted to call a European Summit conference before the end of the year – that is, a gathering of the leaders of EC governments (and the French Head of State). On various occasions during the second part of November, the question was raised of what status exactly this Summit should have and what should be on the agenda. In the French view, discussions should not be limited to the oil crisis. As Pompidou had told Van der Stoel on November 8, priority should be given to the principle question of whether the Nine were actually working toward eventual European integration or not. In addition, the questions of a communal energy and a communal defence policy should also be raised for discussion. Among several matters, the future of the nuclear industry was at issue because of the rivalry between the two competing enrichment plants: the French uranium enrichment concern Eurodif and the British-German-Dutch ultracentrifuge project. The oil crisis, in the French view, should be discussed merely as a part of these wider perspectives.

Furthermore, it was becoming clearer by early December that the French government wanted to use the Summit in Copenhagen to get a dialogue started between the EC and the Arab countries. At the beginning of December, Jobert delivered a speech to the French parliament in which he argued for cooperation and dialogue with the Middle Eastern states. Approaching the Arabs now on the basis of the inconvenience Europe was suffering as a result of the embargo would not go down well in the Middle East. It seemed to him more sensible to try to engage Arab interest in a longer-term dialogue. The French Prime Minister P. Messmer also argued that the oil-consuming states should henceforth themselves come to a direct understanding with the producer countries without the intermediation of the oil companies.[70]

It was clear that Paris thought The Hague had called down the embargo upon itself, and further considered the Dutch energy situation much healthier than had been officially suggested. France's position, as also represented to De Ranitz, was actually more vulnerable, and it was therefore with good reason that France did not want to forfeit good will in the Arab countries. And finally, it was thought that The Hague, more than anyone else, was responsible for blocking any step towards a common energy policy.[71] Taking everything together, France stood directly opposed to The Netherlands on virtually every point. It was still the question whether the other EC countries would support the French outlook and the French plan. France could probably expect most sympathy from London; yet in several respects the French and British viewpoints diverged sharply. London was highly doubtful of plans for a communal EC energy policy. There was also no consensus between the two countries on the question of a common defence policy. Only when it came to a European-Arab political rapprochement was Britain likely to take the French side.

Initially, the Dutch feared that the West German government, and specifically the Chancellor, Willy Brandt, were inclined to share the French orientation, but this began to change during the course of events during November. Brandt's policy began to move more in the Atlantic direction. Differences of opinion between Bonn and Paris were accentuated when the West German Minister of Defence, Leber, emphasized the significance of the Eurogroup within NATO at a meeting of the West European Union. Jobert, on the contrary, preferred a European defence policy independent of NATO, under the aegis of the WEU. Bonn, Leber responded, found the idea of a European defence outside the Atlantic framework unthinkable.[72]

On November 19, in conversation with Van der Stoel, the West German Foreign Minister Scheel commented negatively on Pompidou's proposals. It was still too early to take decisions on such issues as nuclear energy. Under the present circumstances, the first matter was to share the oil shortage on a pro rata basis. Scheel's views no doubt were seized on by Van der Stoel. During this conversation he was not above referring once again to the key role of the Dutch in the oil sector and the West German dependence on natural gas: 'should the need for oil increase and other members of the Community, meaning specifically France and England, refuse to draw the consequences of European solidarity, the pressure in The Netherlands to adopt independent restrictive measures would become irresistible, which would affect the principle of solidarity and lead to a crisis in the Community'.[73]

West Germany was partly dependent on The Netherlands for its energy needs. As Van der Stoel had earlier concluded in a business-like fashion, this was an important reason for West German (and Belgian) willingness to support The Hague. Van der Stoel was probably right. Simonian, the author of a book on French-German relations during this period, also holds the view that the West German change of course was largely the result of the fear of losing Rotterdam as its main transit port.[74]

Belgium and Luxembourg seem also to have responded rather coolly to the French plan. Both governments had been generally supportive of The Hague's struggle for solidarity, despite Belgian irritation with the Dutch attitude, not least because of the threat to restrict the flow of oil and the export of natural gas. Over this latter question, as was remarked in the Dutch Council of Ministers, there had developed what was even referred to as a 'hostile tone'.[75] There was, in addition, another delicate problem between The Netherlands and Belgium. During their visit to Brussels, the two Arab ministers had said they had no objection to unhindered use of the Rotterdam-Antwerp pipeline to supply Belgium with oil. But in such an eventuality the oil would have to be monitored at the beginning of the pipeline, for example by officials from an Arab embassy. As far as this scheme was concerned, however, the Belgians complained that they could expect little cooperation from the Dutch.[76]

Italy seemed a doubtful case. There was undoubtedly a good deal of Italian hostility towards The Netherlands, the *Corriere della Sera* in particular taking a viciously anti-Dutch line in its reporting of The Hague's role in the oil crisis. This paper depicted Van der Stoel's role as that of someone who did not hide his sympathy for the Jews. There were perhaps 200,000 Jews living in The Netherlands, who controlled the financial life of the nation. In spite of this, the Dutch wanted to maintain an existing situation in which the Arabs had to reach the whole of Europe via Rotterdam. For this, they (the Dutch) believed that the EC should show solidarity. The analysis of the daily *Messagero,* however, was that the Dutch government was trying to ensure 'that the multinationals who controlled Rotterdam should continue to be given a free hand' – which was hardly inaccurate. According to the *Corriere della Sera,* The Netherlands, supported by West Germany, had in fact already won the argument with France and Great Britain.[77] The Dutch Ambassador in Rome, Boon, thought that the Italian delegation was unlikely to follow France blindly in Copenhagen. The French effort to get the *states* to negotiate with the Arabs rather than the *majors* would not be shared by Rome. There was far more understanding for the international oil companies in Italy, particularly for Shell and BP, which were seen by the European-oriented as more European companies than as *Seven Sisters*.[78]

Den Uyl and Van der Stoel

From the moment that the French proposals became known, the Cabinet was assailed by doubt. The French plan for a summit conference of governmental leaders was first discussed in the Council of Ministers at the beginning of November, only a day after the proposal had been handed to Den Uyl, who had promised to study it carefully. Den Uyl had further indicated that he was willing to participate. 'Given the Dutch thinking against regularly recurrent European summits, this was an extremely obliging response', was the view subsequently taken at the Foreign Ministry.[79] The Secretary-General of General Affairs (the Prime Minister's Office), D.M. Ringnalda, had meanwhile warned Den Uyl of the Foreign

Ministry's resistance to summit conferences that excluded any foreign ministers. However, Ringnalda thought that every minister was implicated in the collective responsibility for government policy and that, accordingly, the premier could be empowered to defend and prosecute Cabinet policy abroad. It was absolutely not essential that he should always be accompanied by his Foreign Minister.[80]

Predictably, Van der Stoel was in complete disagreement with this view, restating the usual Foreign Ministry standpoint in the ministerial council: a summit threatened to undermine the normal diplomatic cooperation of the EC. He concluded, as he had informed the embassy in Paris, that his presence was necessary on constitutional grounds. But it was not only the nature of the decision-making, the content of the issues raised at the Summit also caused him concern. DGPA Van Lynden had meanwhile stressed to the British Ambassador that in any case he was not much in favour of the idea of taking binding decisions in Copenhagen, which could only be 'impulse' decisions.[81]

On November 16, at the Council of Ministers, Van der Stoel expressed his fear that Copenhagen would be used to make pronouncements directly opposed to American policy. His conclusion was that there was a real dilemma, and he considered approaching Bonn to avert the possibility of damaging Trans-Atlantic harmony. He recognised that there was a proper role for the EC, but The Netherlands should not allow itself to be pushed by that role into opposing Washington. The declaration of November 6 had already caused American-European friction, and any further deepening of an Atlantic conflict had to be avoided.[82] A week later, after discussions on this issue between the EC foreign ministers on November 20, Van der Stoel was able to report to the Council of Ministers that the meeting in Copenhagen was to be only of an informal nature. No communal communiqué would be issued (which actually did happen in the event). A collective gathering of heads of state, government leaders and foreign ministers would be held at the beginning and end of the conference; while between times consultations would take place between government leaders and individual ministers. Van der Stoel did warn that Copenhagen would inevitably have a strongly political character. Issues that would be raised included: the Middle East conflict, the European-American relation and East-West relations, and, moreover, the energy problem, plans for a European Monetary Union (EMU), and possibly even European defence.[83]

This heavily political agenda occasioned great anxiety in the Dutch Foreign Ministry. As he had previously, in the period before the EPC ministerial session of November 5 and 6, Van der Stoel hoped that the Copenhagen Summit would concern itself with the oil problem as much as possible and attend to the political objectives advanced by the French as little as possible.

The idea was to prepare for the Summit during an EC Energy Council meeting over December 3 and 4, but the Dutch were from the beginning pessimistic about this meeting. During a bilateral Ministry of Economics – Foreign Ministry meeting, Lubbers also let it be known that he expected little from the Energy Council. He confirmed that the oil companies were under pressure from some governments, but Cabinet action seemed unnecessary. He did have the impression that the companies would not be able to keep up the game of sharing out the oil (by shifting oil originally from Nigeria and Iran).

> At the moment, some Arab oil is arriving in Rotterdam (which is apparent from the decreasing imports in England from Nigeria), but it is unclear how much is being transferred. Of course, we are exercising maximum reticence over this.[84]

That pessimism seemed to be confirmed when it was learned that the European Commission had worked out two new proposals for implementing an oil allocation arrangement and for setting up an advisory committee for the oil industries. London and Paris reacted angrily to this plan, which Van der Stoel read as a signal that both countries 'found the speculative chance of a new political role in the Middle East more important than optimal European cooperation'. Preparatory to the EC Energy Council meeting of December 3 and 4, which was also taking place in Copenhagen, Heath again stressed that the EC should concentrate on the foreign policy aspects of the Middle East conflict.[85]

During the EC Energy Council of December 3 and 4 in Copenhagen, attended by Ministers of Foreign Affairs, Economic Affairs and Finance, the differences of outlook within the EC were once again clearly delineated. Van der Stoel again set out his standpoint emphatically. He pointed out that The Netherlands was now beginning seriously to feel the restrictions, as a result of which it was the first EC country to announce petrol rationing (see Chapter 6). Van der Stoel was critical of those who were only interested in the political side of the Middle East conflict, meaning first and foremost Great Britain. It was an illusion, however, even if some magical formula for a speedy resolution of the Middle East conflict should appear, to think that the embargo and restricted production would be lifted within a few weeks. And their effects, moreover, were likely to last for some time. There was no invisible hand ensuring equal shares of the oil supply to all member states. Should the Nine be incapable of taking any (discrete) communal action, the market would disintegrate with all the consequences this implied. The only just and proper step, according to Van der Stoel was joint, communal action.

Jobert replied that sharing the poverty would be no solution at all. And moreover, it was quite unjust to point to France and Great Britain as privileged countries. They were absolutely not energy-rich countries; France was in fact poor in energy reserves. Paris had long tried to promote the creation of an energy policy and to build up stocks, repeated Jobert, implicitly referring to The Netherlands. In this context, one must speak of past, present and future, and over the energy question as a whole, i.e. including the question of uranium enrichment, which equally called for solidarity. Addressing Van der Stoel directly, Jobert suggested that panic and psychosis had to be avoided if relations with the Arabs were not to be made even more difficult.

The British Minister for Energy repeated his position, that openly communal action would endanger oil supplies by provoking Arab counteraction. In the UK too, preparations for rationing were in hand, and joint diplomatic action, in his view, offered better prospects. The Arabs had to be convinced that the EC would not be undermined.

Support for Van der Stoel came mainly from the Belgian and the West German side. The Belgian Foreign Minister Van Elslande wondered what would remain of the Community's credibility if the EC was incapable of resolving the oil problem on a communal basis. The West Germans also favoured communal action. To the West German mind, the coming Copenhagen Summit should begin with discussion of current energy problems. These views were also endorsed by the Danish and Luxembourg side. The Luxembourg Foreign Minister Thorn said he would find it rather schizophrenic if the Nine should adopt a communal political standpoint toward the outer world whilst not recognising the EC as an economic entity. The Chairman of the European Commission, Ortoli, also supported the Dutch view.[86]

Taken together, what this meant was that The Hague's views on the necessity for communal action to deal with the reduction in the oil supply were by now shared by a considerable number of EC member governments. In view of the powerful opposition of both Paris and London, however, any movement by the EC or the European Commission was ruled out. On the contrary, the Commission's proposals were brushed aside. Nor did it look as though these would be dealt with at the Copenhagen Summit to be held in ten days' time. An embittered Van der Stoel declared after the meeting of December 3 and 4 that the EC member states would have to suffer the consequences themselves if the Dutch economy were radically affected by the oil crisis.[87]

Visible or Invisible

In The Hague, a growing sense of doubt over the relationships within the EC gradually came to dominate the outlook.[88] It was very much open to question whether the EC partners could ever agree on any communal plan of action to deal with the oil crisis. But in addition, it was becoming clear that The Netherlands was perhaps not in such a bad situation as had been assumed throughout November. In the Ministerial Council for Economic Affairs, it was now concluded that, in view of the relatively satisfactory Dutch position, (communal EC) *sharing* carried considerable risks as well, certainly if natural gas was also involved.[89] As long as the oil companies shared out the oil pro rata, The Hague was actually in rather good shape.

This is not to say that the December 14 and 15 Summit was not thoroughly prepared in The Hague, notably on December 6 in the Council for European Affairs and the following day in the Council of Ministers, on the basis of two notes prepared by the Foreign Ministry and Ministry of Economic Affairs, respectively. The Foreign Affairs note put forward the view that regular summits were undesirable, because they would interfere with the primary responsibility of the Ministry of Foreign Affairs. The same note recognised the major importance of communal political action, but noted the undesirability of London and Paris being given an EC mandate in any form. The note also rejected the idea of participation of either country at the coming peace conference in Geneva, as well as EC pressure being brought to bear in order to achieve that objective. Such participation would arouse great opposition in Israel, which could only serve to delay any peace accord. The note also rejected the idea of a new declaration by the Nine. Kissinger's standpoint was rather that it was now time for *quiet diplomacy*. In general, in view of the importance of a healthy Atlantic alliance for the security of the West, as well as the sound financial-economic relations of the world, it was essential to be alert to anything that might try to define a European identity in opposition to the USA.[90]

With regard to the oil problem, the note proposed that discussion of this issue should have absolute priority at the Summit.

> Under the present circumstances, it is clearly of the utmost importance to The Netherlands that at Copenhagen, in some shape or form, there should be evidence provided of an awareness of communal responsibility for the solution of this problem.[91]

In his clarification of this note, Van der Stoel concluded that the central question was whether the Summit was prepared to make the oil problem a communal issue.

The minister sketched the positions within the EC regarding this question as follows: powerful support could be expected for The Hague's viewpoint from the West German Republic, Denmark and (to a lesser extent) Belgium and Luxembourg, but the attitude of Great Britain, France and Italy gave little cause for joy. He hoped that it would be possible to deal with the oil problem 'above all other topics', but he was not optimistic.

The members of the Cabinet appeared to agree with the Foreign Ministry note and Van der Stoel's elaboration of it. Den Uyl, Pronk and Finance Minister Duisenberg merely queried the passage repudiating the institutionalisation of summits in general: in Den Uyl's view, such meetings often had played a stimulating role. Van der Stoel replied that summit conferences threatened to undermine the work of both the Council and the Commission. Their regular occurrence should therefore not be endorsed.[92]

Lubbers's note, drawn up in consultation with the Foreign Ministry, was equally pessimistic over the possibility of Copenhagen reaching a communal policy regarding the oil problem. It dealt extensively with the plans for a communal EC energy policy, which was regarded with some caution by the Ministry of Economic Affairs. It was clear, said Lubbers, that Paris and The Hague took very different positions. The Netherlands, he said, had now made its standpoint a more flexible one, but there was no reason to take the lead. The Netherlands was ready to cooperate in bringing about a community price policy. The communal promotion of research was also significant. It was most important that a central role should be reserved for the European Commission, whose outlook was totally at variance with the French. Lubbers was therefore apprehensive that more far-reaching decisions might be taken in Copenhagen, for example over the basis of a communal energy policy, or over nuclear energy, while providing no guarantees for the communal operation of the energy market.[93]

During the debate in the Council for European Affairs, differences of opinion seemed to be developing along party political lines. According to Pronk (PvdA), the Ministry of Economic Affairs note was excessively preoccupied with state interference. Den Uyl expressed himself more cautiously, but thought there was no sense in leaving everything exactly as it was; for one would then be entirely in the hands of the oil companies. In this context it was good to remember Edward Heath's behaviour at the start of the oil crisis. Den Uyl also wanted to know why The Netherlands had for so long resisted a communal energy policy. In response to this question, the acting Director-General for Energy from the Ministry of Economic Affairs, A.T.T. van Rhijn, said that The Netherlands had obstructed such a policy because of its natural gas reserves and because of Rotterdam's special position. Moreover, it was feared that France, whose oil politics were closely tied to foreign policy, would be able to push the EC into a Middle Eastern policy that was unacceptable to The Netherlands.

A last, striking aspect of the debate in the Council for European Affairs was the question of natural gas. Lubbers declared that, domestically, oil was increasingly being replaced by natural gas, which had its repercussions on export. There should be no reduction in exports because of this, but there should be a deceleration of export growth. He wondered why this could not be said publicly. Not speaking about this would only lead to speculation over The Hague's plans for natural gas. In the end, one ran the risk of political arrangements over the inviolability of natural gas contracts, both in relation to volumes and prices.

On this point, as so often, Van der Stoel was more cautious. The first priority was solidarity. It was scarcely possible to demand solidarity and at the same time threaten other partners with export restrictions. And in the end, such an action would predominantly affect precisely those countries that supported The Netherlands. Den Uyl appeared to be more Machiavellian than his party colleague. He thought that one should paint the nation's situation as it was, without threats. There was nothing improper in taking up a position openly.[94]

The following day, this exchange of views was largely repeated and reaffirmed in ministerial council. The Council decided in principle to go along with a communal energy policy together with research and development. In the first place, however, the question was about taking communal action to combat the oil shortage, and above all directed at bringing down the level of consumption. The chances of success were not estimated highly. Van der Stoel pointed out that London was on the whole not in favour of a communal energy policy. He once again warned of the French plans for a Euro-Arab dialogue. France was working hard for this, using other means than purely through discussions between producers and consumers. Such a dialogue would create serious friction with Washington, while Kissinger undoubtedly would totally reject any thwarting of his political aims in the region.[95]

Van der Stoel's expectations of Copenhagen remained low. On December 4 he had tried to convince his Belgian and Luxembourg counterparts that a common Benelux standpoint would be useful, referring to the danger of the French plans for a Euro-Arab conference, specifically with an eye to American-European relations. Van Elslande was not convinced by Van der Stoel's argument and became more positive in his attitude to the French ideas.[96]

Attempts at consultation with Paris and London also failed to deliver. De Ranitz reported that urging '*solidarité de fait*' had produced no result. The French attitude toward The Netherlands in Copenhagen would be a hard one: if the Dutch were unwilling to cooperate on the French plans, they would themselves be responsible for their own oil problems. Shortly before the conference, it became apparent, not for the first time, that there was no support to be expected from the British side. Van der Stoel had emphatically put the question to Home, the British Foreign Minister; while the following day Ambassassador Gevers reported a conversation with an Assistant Under-secretary of the Foreign Office. Open expressions of solidarity with The Netherlands, according to this British diplomat, were counterproductive. Furthermore, The Hague was at the time no worse off than its other EC partners, thanks to the operations of an '*invisible hand*'. Why then demand a '*visible* hand' (i.e. 'sharing' between member states) that could not itself produce any oil to share out, when the result would be merely to restrict the activities of the '*invisible* hand'.

This was a view shared by Shell. The oil companies were well prepared to take care of a fair share-out of oil with their 'invisible hand'. Wagner was also unconcerned about possible practical problems. His advice to Van der Stoel at the Copenhagen Summit was 'to play it in such a way that the Dutch should for the time being be content with a *solidarité de fait* and should keep this as much as possible to themselves'. Should the British and the French proceed with their restrictive practices and should they, for instance, announce formal unilateral measures, The Hague could always resume its own freedom of trade.[97]

The *visible hand* was therefore wholly unnecessary; The Netherlands was in good shape. Self-confidence in The Hague received a further boost when, on December 9, on the eve of the NATO conference of ministers, Van der Stoel met Kissinger in Brussels. Kissinger expressed wholehearted appreciation of the Dutch attitude, and not without reason. On different occasions, The Hague had tried to act as mediator between Washington and the Nine. On November 28, the Dutch DGPA Van Lynden had assured the American ambassadorial counsellor that The Hague would do its best within the framework of the European Political Cooperation to exercise a moderating influence. Van der Stoel's meeting went very well, despite Kissinger's initial distrust of his socialist opposite number. In the event, the two ministers agreed on almost everything. Kissinger objected strongly to the construction of a European identity on the basis of anti-Americanism. Van der Stoel assured him that European solidarity, in The Hague's view at least, must always be placed in an Atlantic context. Of course, Van der Stoel was curious to know what Kissinger thought of the embargo. Kissinger's response sounded highly promising. He let it be known in strictest confidence that Yamani had promised him in fact to try to get the embargo against The Netherlands lifted.[98]

Van der Stoel also adopted the position at the NATO ministerial conference that one would expect of an Atlantic statesman. The meeting was distinguished mainly by the sharp confrontation between Jobert and Kissinger. Kissinger did not hide his irritation over the attitude of the Community, while Jobert argued precisely for more drastic, independent European moves in the Middle East conflict. Van der Stoel tried to support Kissinger during the conference, which ended finally in a much more pleasant atmosphere than it had begun. Certainly, there was no hint of conflict in the final communiqué. However, Van der Stoel had found himself in an awkward predicament, given that The Netherlands had constantly exerted itself to prevent European-American estrangement. It was an important effort for the Dutch, since 'The Netherlands more than any other EC country saw the maintenance of the link between the two continents as an important goal of its foreign policy'.[99]

At the close of the conference, a conversation with his Benelux colleagues left Van der Stoel with an even more despondent outlook on the coming Copenhagen Summit than before. His gloomy reflections were conveyed to the Dutch Foreign Ministry:

> A working breakfast with Benelux ministers (plus Directors-General) ... led to an extremely depressing prognosis for the coming summit, where the French want to take all the decisions (in league with the UK, but hopefully opposed by West Germany) through a series of *faits accomplis*, largely setting aside or circumventing community institutions and procedures.

The French proposals were revealed the same day. One striking passage was the suggestion of creating a European Security Council, possibly comprising Under-Secretaries of State, which could meet at any desired moment in crisis situations. From Paris, De Ranitz found these proposals remarkably similar to the *Fouchet plan* of the early 1960s, plans against which The Netherlands, under the leadership of Minister Luns, had fought tooth and nail at that time. Nonetheless, it hardly seemed an attractive prospect to the Dutch that they would be made scapegoats if the proposals were not accepted by the Summit.[100]

Meanwhile, it was also learned that several Arab ministers were coming to Copenhagen for consultations with European leaders. It seemed highly likely that this visit had been arranged at French instigation. Such a meeting was totally unacceptable as long as the embargo was in place. If the talks with the Arabs were to go ahead, Van der Stoel demanded as preconditions that discussion should first be held between the Nine over the manner of their reception; that there must be no discrimination between the EC partners; and that any real exchange of views with the Arab ministers must be avoided. In any case, the peace conference in Geneva must not be frustrated. These conditions were handed to the Belgian and French governments.[101]

Matters were complicated still further on the eve of the Summit when Kissinger, in an address in London on December 12, dealt at length with the oil crisis. He called on OECD countries to develop joint initiatives regarding oil production, a more rational use of energy and the development of alternative energy sources. Kissinger proposed setting up an *energy action group* to discuss these proposals further (see Chapter 7). Van der Stoel responded immediately and positively to these proposals, which were of course intended to influence the Copenhagen Summit and to stymie any European go-it-alone, such as the French were trying to promote. The French daily *Le Monde* branded the speech as a crude meddling in European affairs.[102]

Copenhagen

The Summit began on December 14. The afternoon saw the arrival in Copenhagen of government leaders and foreign ministers. Preliminary talks had been arranged for the first evening, followed by the arrival – highly undesirable in the Dutch view – of the foreign ministers of Algeria, Tunisia, Sudan and the United Arab Emirates. During conversation, the four expressed their hope that the EC would play an active role in the coming peace process and that this should not be left to the two superpowers. The EC should insist on Israeli withdrawal from all the occupied territories and on respect for Palestinian rights. The Arab countries wanted closer cooperation in general with Europe, in economic, technical and cultural fields. At the same time, it was not expected that Europe would turn against Israel, but it was hoped that the future would bring greater activity in favour of the Arabs.

The Nine, however, at least a majority of the Nine, appeared unwilling to meet the Arab delegation. As recorded in a report from Van der Stoel, it was decided that the Danish Prime Minister, A. Jörgensen, and Foreign Minister, K.B. Andersen, would receive the Arabs once more, 'so that aside from courtesies, the damage caused by the Arab actions to the European economies could be discussed'. Promises were given to look further at the Arab wishes, but that was all.[103]

Pompidou had originally expressed his preference for further talks with the Arab ministers and wanted the Nine to work out a mutually agreed standpoint over the Israeli withdrawal from the occupied territories. But it was Brandt, in particular, who insisted that the declaration of November 6 was absolutely sufficient and that with regard to resolution 242 Europe must not be more pro-Arab than the Soviet Union.[104] The Italian Prime Minister also stressed that negotiations with the Arabs should be avoided. Den Uyl naturally endorsed the West German standpoint. The further talks that Pompidou wanted did not happen.

The atmosphere was not improved when it was suggested by some – specifically the smaller – member states that the Arab visitors had been secretly invited by Paris and possibly also by London.[105] In the event, however, the French delegation did not succeed in notching up a single victory over the Middle East conflict. West Germany, The Netherlands and Denmark forcefully resisted a French proposal regarding the evacuation of the Israeli occupied territories. A rather milder text from the British was also seen off. In the end, the French President proposed expressing the hope that Israel realised the significance of the fact that the Arab countries had accepted peace negotiations. Den Uyl resisted this proposal too, because it could be construed as a one-sided appeal to Israel. With support from Brandt, pre-eminently, and from the Danish Prime Minister, Jörgensen, and his Foreign Minister, Andersen, this text was also struck off.[106]

Pompidou thus achieved none of his aims, which was partly explained by his poor health. It struck Den Uyl that Pompidou looked ill, 'completely swollen up by cortisones', according to the Dutch Premier. It was for this reason that Pompidou was incapable of standing long meetings. His poor condition was confirmed by the Danish Prime Minister, Jörgensen, who noted in his diary that the French President was on the verge of collapse, with an ambulance waiting permanently at the ready.[107] The most important reason for the French failure, however, was the growth of opposition during the preceding weeks, notably that of the West Germans. In the end, the press communiqué contained no more than an appeal to reach a just and lasting peace accord, referring to the declaration of November 6 and resolution 242 in all its parts.

Pompidou refused to enter into discussion of the current oil problem, partly in reaction to the preceding attitude of the majority of member states. He declared that this question should only be raised as an integral part of discussion of a communal EC energy policy. But it was very quickly evident that there was no agreement over the communal, more US-independent EC energy policy being pushed by the French. Only Heath, who during the dinner had called Shell 'a curse', was ready to support the extreme French proposals, although it was open to question whether London was really prepared to cooperate with a communal energy policy. The other member states turned against the French proposals, thus opening the way for the discussions between consumer countries that Kissinger was arguing for. In the end, it was decided to ask the Council of Ministers and the European Commission to develop proposals and to take appropriate action toward establishing a communal EC approach, especially to the current oil and energy problems.[108] From the very beginning, the question had been whether this would actually happen.

All in all, Copenhagen had produced more conflict than cooperation, and for this very reason had delivered a result that was wholly acceptable to The Netherlands. The Euro-Arab dialogue, for the time being, was a non-starter. The radical French plans for a communal EC energy policy had only been discussed in the most general terms and had led to nothing in the way of decisions. Nor had anything concrete been arranged over nuclear energy. The American proposals for consumer discussions had not been rejected, and, furthermore, the final communiqué referred to the need for Atlantic unity. The possible institutionalisation of summit conferences of government leaders was mentioned, but the idea of a European security council was turned down, specifically by Willy Brandt. France and Great Britain, in fact, had suffered a defeat, while the Dutch position appeared considerably strengthened. Van der Stoel subsequently expressed his satisfaction over the Summit.[109]

Another feature of these events that had become clearly apparent was further estrangement between the Germans and the French, which also suited The Hague. This was in part caused by the fact that Bonn was not prepared to go on financing a large part of the joint community agricultural policy, which favoured the French.[110] For similar reasons, this time to do with regional support given to Great Britain, British-German relations had also become rougher.

There still remained a few unpleasant aspects of the Copenhagen Summit. In all probability leaked by the French, reports appeared in the press to the effect that particularly Bonn and The Hague had resisted a more critical statement on the Israeli position in the Middle East conflict. The content of the French draft text, which in the event was not accepted, was also published before the final communiqué. The resentment which this caused was brought to the attention of the French Ambassador by the Dutch DGPA Van Lynden.[111]

But by then this could not harm The Hague. Two days after Copenhagen the postponed visit of American Secretary of State Donaldson took place. Lengthy discussions were held with him in the presence of Ambassador Gould. The first session of these discussions was led by Van der Stoel, who expressed his satisfaction with the outcome of the Summit, since the principle of community action had now finally been accepted. At the same time he stressed the importance of consultation with other oil-using countries, specifically the USA. From the American side, admiration was expressed for the Dutch Cabinet's attitude and its refusal to be blackmailed by oil, so unlike other European countries. On the other hand, there was disappointment at the fact that the final communiqué omitted any explicit reference to the Kissinger plan. Gould and Donaldson made it clear that the aim of their visit was to be informed of the situation in the energy sphere. It should therefore be said at once that any American possibilities for lending assistance were limited. It was also not the intention to put The Netherlands into a difficult position within the EC by lending help.

At the second session, led by Lubbers's the main topic of discussion was the practical situation. Lubbers explained that the government was in the meantime busy saving over the odds on petrol consumption in order to be able to pass on sufficient oil products, specifically naphtha, to the chemical industry. He enquired about the possibilities of American oil being supplied to Rotterdam refineries from American companies, about the possible supply of American coal to the Hoogovens steel works and about the use of NATO oil stocks by the Dutch armed forces. All this would have to be looked into in Washington.

These discussions would have no immediate, material, sequel. It was more a matter of political and psychological gesture, a helping hand. A brief account was distributed to the British and the French Ambassadors stressing that The Netherlands preferred solutions within the context of the EC and OECD, and that if there were any American help with oil there would be no conditions attached.

Conclusion

In various ways, Copenhagen can be seen to have been a failure. The route to a communal energy policy was blocked; and no concrete, communal approach to the consequences of the oil shortages was achieved. Dutch politicians at The Hague publicly spoke of their great disappointment. Press opinion over the Copenhagen Summit was also bitter. 'Europe – does it exist?' *Het Parool* wondered. When needed, 'the Brussels machinery was full of the sand of nationalism', concluded the Amsterdam newspaper.[112] *Trouw* found it astonishing that, 22 years after the institution of the European Community for Coal and Steel, Western Europe 'was still toiling over the creation of a communal energy market'.[113] Other dailies also remarked on the total absence of any perceivable 'communal solidarity' within the EC.[114]

Yet more privately in The Hague, the feeling was mainly one of satisfaction at the outcome of Copenhagen. In PvdA circles, too, the Summit was seen as a Dutch success. This is very evident from a note written by Harry van den Bergh, Secretary of the PvdA EC Committee.

> Though practically ignored in the Dutch press (in contrast, for example, to the English press), one can safely say we are looking here at an important political and diplomatic success for this Cabinet.

According to Van den Bergh, the two most conservative governments in Europe, France and the UK, found themselves opposed by a united front of The Netherlands, West Germany, Denmark, Belgium, Ireland and Italy. Pompidou had not come to Copenhagen to talk about the energy crisis. At the very outside, he would have been prepared to deal with this in order to coerce the other member states, apart from Britain, into accepting the French political line. 'Yet one must suppose that, as far as this is concerned, the planning in Paris for Copenhagen went somewhat astray', he observed. The final communiqué from Copenhagen reflected the lack of results achieved by the French and British.[115]

The Netherlands had rather effectively contributed to this outcome. It had throughout forcefully resisted the French and British standpoint, both the political and economic aspects. Of course, it was mainly the powerful attitude of West Germany that had defeated the French and British proposals, but it is not entirely unlikely that the West German dependence on oil supplies from Rotterdam and on Dutch natural gas had contributed to the West German point of view. In any case, Lubbers and Van der Stoel had on several occasions referred to this dependence.

No definite arrangements were made over any communal approach to dealing with the oil shortage, for which the Cabinet and Van der Stoel had so trenchantly fought. But was that in fact such a disaster? By that time, the Cabinet had come to see clearly that The Netherlands stood to gain little from such a policy. On the contrary, The Netherlands might well find itself worse off if oil provisions were in some form or other equalised within the EC. The *invisible hand* of the oil companies was highly generous to The Netherlands, as Wagner had emphasized to Van der Stoel. It was therefore essential to prevent a situation within the EC where member states would try, through legal means for example, to hold the oil companies to their contracts. As long as that could be avoided, The Hague's position was actually not bad. Furthermore, there was the offer, albeit a rather vague one, of American support. All in all, The Netherlands stood to suffer under the Arab actions no more – and possibly a good deal less – than the other EC countries. Added to which, the Dutch commanded considerably larger reserves of their own energy in the form of natural gas.

Van der Stoel had by now come to the conclusion that making political concessions to the Arab countries was not going to produce results. Steps taken towards the Arabs had so far led nowhere; the outcome of the new EC declaration of November 6 had been disappointing. The meeting Lubbers had held with the two Arab ministers had similarly produced nothing. The Hague, moreover, had been pushed into a point of view that seemed scarcely different from that of most other EC countries. Without ever explicitly admitting it, the Den Uyl Cabinet had also adopted the standpoint that Israel must vacate *all* the occupied territories. Only in The Hague must this not be said in so many words. When Foreign Ministry spokesman Thurkow lost sight of this tacit understanding, he also lost his position. The time for explicit support for Israel at the beginning of the war was past. Nor was it necessary, since Israel had emerged from the conflict as victor. The areas under Israeli control had enormously expanded, although this also meant much greater sacrifices.

None of this, however, much altered the fact that The Netherlands, along with the other EC countries, was confronted with production restrictions. In the next chapter, we should therefore pay some attention to the domestic measures introduced at this stage to reduce domestic energy consumption.

6
Rationing

In this chapter we shall turn our attention again to the policy adopted by the Dutch Cabinet to compensate for the reduction in the oil supply. We pick up the thread early in November, when the first restrictive measures were introduced to limit oil use. Throughout the course of November, assessments in The Hague of the consequences of the Arab oil actions became increasingly gloomy. This pessimism reached its peak at the end of the month when the possibility of a future reduction in the oil supply of some 40 to 50% was being discussed, even at the level of the Council of Ministers. In addition, the Dutch Central Planning Bureau was predicting that the oil embargo would lead to growing unemployment and to inflation.

Other noises were also being heard: the view was taking root in various quarters that maybe The Netherlands was not in such a bad state after all, certainly when compared with most other EC countries. Yet nobody could say with any certainty how the supply would develop, and for this reason the dominant general feeling was that radical measures would have to be taken to compensate for the effects of the oil embargo. It was in this phase of uncertainty that the Cabinet took the decision to set in motion the preparations for rationing.

At this stage, the Dutch Cabinet not only decided on rationing but also submitted to Parliament the so-called Enabling Act. This law authorised the government to take socio-economic measures on a sweeping scale in order to cope with the crisis situation caused by the embargo. With rationing and the Enabling Act, the political-economic heavy artillery was in place. It was also at this stage that the Ministry of Economic Affairs lost its leading role in the introduction of measures to limit consumption, as combating the effects of the embargo became increasingly an affair of the entire Cabinet, led by its Minister-President.

Car-Free Sundays

As we saw in Chapter 4, several authorities were called into being at the beginning of November to deal with the consequences of the oil crisis, and specifically with a reduction in the domestic use of energy. The most important official body was the Co-ordination Group, consisting of civil servants from various Dutch ministries under the leadership of the Secretary-General of the Ministry for Economic Affairs, F.W. Rutten. The Co-ordination Group played an important role during November and December as the supplier of plans, data and decisions to the Council of Ministers, where the group's suggestions were in many cases taken on board. Although various ministries were represented in the Co-ordination Group, Economic Affairs initially took the lead in developing measures to limit consumption. The Director-General for Energy and his deputy Director-General Van Rhijn had enormous influence on Economic Affairs' policy-making. The preparation of policy at Economic Affairs was to an important extent co-ordinated by the Oil Crisis Work Group.

Economic Affairs provided the Co-ordination Group and the Council of Ministers with information on the supply of oil and oil stocks, with regular meetings taking place between Economic Affairs and the oil companies, at different levels and in different contexts. For instance, consultations were arranged within the so-called Oil Contact Committee, where representatives of Economic Affairs and participants from the world of the oil companies held sessions together. As we saw earlier, provision of information on supplies, stocks and estimates by the companies had been raised to a new level since the beginning of the crisis. As far as can be judged from the documents, the policy advocated by Economic Affairs was generally supported by the oil companies. In fact, it sometimes seemed as though Economic Affairs served as the mouthpiece for the main oil multinationals; so much so that within the Council of Ministers the figures presented by Lubbers and his Ministry were sometimes queried.

The most significant of the measures enacted to restrict oil use was the car-free Sunday, introduced at the beginning of November. At the same time, the Dutch people were called on to cut back on their use of energy. On November 14 it was confirmed in the Co-ordination Group that the car-free Sunday resulted in a drop in petrol consumption of around 10%. Furthermore, roughly 90% of motorists had observed the voluntary speed limit, leading to a further reduction of 5% in petrol consumption. In all, a total reduction of some 15% had thus been achieved. The oil companies wanted Economic Affairs to limit the production of various products – petrol, gas for cars, diesel, paraffin and fuel oil.[1]

Nevertheless, the dominant feeling in early November was that the car-free Sunday and the speed restrictions would not be enough: further measures were going to be needed. At its very first session, the Co-ordination Group discussed various other possibilities, such as banning driving on other days beside Sundays. This kind of alternative invariably provoked numerous objections. A driving ban on weekdays would lead to total dislocation of the nation's economic and social life, it was felt. A driving ban on Saturdays would mean an extra blow for the catering industry, which had already been badly affected. The ban on Sundays could possibly be extended to three o'clock on Monday morning, but that had to be the limit. A driving ban over the Christmas holidays was also rejected.

A legally enforced maximum speed limit was seen as a possibility; the legal implications of this possibility would have to be thoroughly looked into. There was apparently little confidence in the Dutch motorists' willingness to hold voluntarily to a 100 kms per hour speed limit over the long term. During the following week, ideas were exchanged between various bodies over this legal speed restriction, including even the Council of Ministers. During the session of November 9, it was already evident that the Ministry of Justice had objections, not least the limited possibility of enforcing such a speed limit. Lubbers therefore argued for creating a special law, to which end talks would need to be held between the Ministries of Justice, Transport and Water Management, and Economic Affairs. The Council accepted Lubbers's proposal, but in the end, and despite all the consultations, no such emergency law was introduced.

Throughout this discussion and indeed from the outset, it had been realised within Economic Affairs and in the Co-ordination Group that, should the worst come to the worst, measures like reducing the speed limit were hardly going to be of much use. The only truly effective way of reducing consumption was rationing; and indeed far greater reductions could then be achieved. Rationing would allow the government to have its own hand actually on the oil tap. The great advantage, moreover, was that all driving prohibitions could be dispensed with. It would become unnecessary to prohibit road traffic either partially or wholly on any particular days, with all the consequences of such bans; motorists would themselves decide how and when they should use their scarce resources.[2]

On November 9 the suggestions of the Co-ordination Group were discussed in the Council of Ministers. For the first time, the ministers seriously took on board the possible implementation of rationing, with Den

Uyl as advocate. Should the embargo persist, he said, rationing would be inevitable. Most of the others were of the same mind. The decision was therefore taken to put in motion all the necessary preparations for rationing, using vouchers, although the final decision to actually implement rationing was *not* yet taken. Effectively, however, the decision set in motion a bureaucratic machine that could no longer be stopped.

The Dutch Council of Ministers also agreed to a rise in the price of petrol and other oil products. Lubbers told the Council that 'with heavy heart' he had approved the price rise for oil products, necessitated by the higher cost of crude oil, the increase in the price of petrol being 2.5%. It was conceivable, he thought, that some of the *majors* might be taking advantage of this, because of the fact that they still held stocks bought in at the earlier, cheaper, price; but this was no reason to postpone the decision.[3] In the following months the price of oil products was to rise still further.

During the course of November, the introduction of rationing became increasingly inevitable as alternative methods of reducing petrol consumption were rejected by the Cabinet. The Council of Ministers, following the conclusions of the Co-ordination Group, found it difficult to extend the system of banning car travel. Car-free holidays were unacceptable, in view of the radical consequences this would have for family life, as well as for the catering industry. After long debate it was decided to maintain the Sunday motoring ban that was due on December 2, notwithstanding the approach of the feast of *Sint Nicolaas* on December 5. A week later, it was decided to extend the car-free Sunday to 3 AM Monday morning, as the Co-ordination Group had proposed.[4]

Meanwhile, the first consequences of the Arab actions were felt in the port of Rotterdam. The predictions over supplies of oil in the near future were more pessimistic than ever. Against this background, the Dutch Cabinet decided on November 23 to initiate the concrete arrangements needed for introducing ration vouchers.

Estimates

By mid-November, estimates in The Hague of future reductions in the oil supply and their consequences were grim. Although it was widely established that the actual situation was still satisfactory, it was mainly the *anticipation* of shortages in the coming months that became more and more pessimistic. This pessimism was in the first place propagated by the Ministry of Economic Affairs on the basis of data and expectations communicated to them by the oil companies. In a meeting of the Economic Affairs Oil Crisis Work Group on November 21, it was assumed that it would be necessary to reduce petrol consumption by 20% in December, and 40% in January.[5] Obviously, the car-free Sundays and voluntary speed restrictions would be inadequate to achieve targets of that order.

These gloomy predictions originated with the oil companies, and they were the main compelling reasons for switching to rationing. In retrospect, it can be seen that these estimates for December and January were, to say the least, on the bleak side. An ex-director of Shell-Netherlands, A.C. Helfrich, acknowledges that these estimates from the oil companies were rather 'conservative'. One of the factors affecting these estimates was the fear that the American market might suffer shortages and exert its enormous pulling power on available oil supplies.

According to Helfrich, Shell was opposed to rationing.[6] But from the minutes of the Co-ordination Group, it would appear that the oil companies were at any rate advocating further restrictions on consumption beyond the 10 to 15% maintained thus far.[7] This standpoint inevitably implied rationing. The companies were thus partly responsible for setting in motion a series of measures which would, in the new year, finally lead to rationing being introduced. In general, as ex-minister Westerterp later commented, one could still talk of close consultation between government and oil companies, and the companies supported what the government was doing, including the introduction of rationing.[8]

On November 23, the Dutch Council of Ministers also learned of the pessimistic forecasts of the Ministry of Economic Affairs' Oil Crisis Work Group and the Co-ordination Group, which, as Den Uyl reported, were confirmed by a letter from Shell's Chief Executive, Wagner. The tone of this letter was sombre. Wagner thought the rapid, drastic price increases would have a disastrous effect on the chemical industry and that the whole world economy would suffer. At that time, American oil companies were already buying Nigerian oil at prices three or four times the normal price of oil. The price of petrol in The Netherlands, Wagner thought, would shortly have to go up by some 20%.[9]

These gloomy prognoses were to play a principal role during the coming weeks in decision-making by the Council of Ministers. The Central Planning Bureau added its own voice to the chorus of pessimism. On November 7 Lubbers, with the approval of the Council of Ministers, had asked the Central Economic Committee for their advice on the initial global consequences of the oil embargo. This advice, which was based on a note of November 15 from the Central Planning Bureau, was sent to Lubbers at the beginning of December. It painted a sombre picture. It was assumed that the oil supply in January would be about 50% lower than in 1972, leading to little or no economic growth, increased inflation and rapidly rising unemployment. The Central Economic Committee concluded from these figures that the government had to implement a string of measures if they were to keep in check the consequences of the embargo and the sharp rise in oil prices. The note would play an important role in the tabling of the Enabling Act and in persevering with the voucher system of rationing.

Economic Affairs was also assuming at the end of November that the oil supply in December would fall by around 35%, and even by 50% in January. It was widely thought that the shortfall during the subsequent months would remain at about 30 to 35%. By eating into existing oil stocks, the necessary reduction in the domestic consumption of oil products could be held at around 25%; but in this case it would be necessary to give priority to the supply of energy and raw materials needed for trade and industry. At a meeting of the Oil Crisis Work Group in Economic Affairs, the conclusion was drawn that petrol rationing would have to lead to a 40% reduction in the use of private vehicles, and a reduction in business traffic of 20 to 25%. The total reduction in petrol consumption would have to amount to approximately 30% if sufficient freedom were to be maintained for trade and industry.[10]

Gloom over the immediate future reached its deepest point in the Council of Ministers on November 30. Lubbers informed his colleagues that the beginning of a reduction in supplies had been confirmed in Rotterdam. In December, the reduction in supply would reach 60% compared with the previous year; while the figure should be 50% in January, he said. This would probably mean a drastic cutback of production at the refineries. The oil stocks would have to be opened. Above all, the drop in production of naphtha could cause problems for the chemical industry. Lubbers therefore wanted to maintain naphtha production as far as possible at an optimal level, which would mean cutting back on petrol for road transport. By mid-April or May of 1974, stocks could be expected to have fallen to a reserve sufficient for only 40 to 45 days.

Lubbers said the reduction in supplies could in the longer term amount to as much as 60%, since two-thirds of the total supply came from countries enforcing the embargo. Because of the increased supply from Nigeria and Iran, the total drop now was around 50%. In comparison, other West European countries were confronted with reductions of around

25%, and for this reason Lubbers thought that Dutch exports could be reduced without causing serious problems to other EC countries. In council, he urged once again to speed up the introduction of rationing, though he realised that it would not be possible to achieve this in December. Both the oil companies and the consumers were pressing the government to set out a system of priorities. If the Council of Ministers were to decide to begin rationing on January 7, he wanted the power to be able to instruct the oil companies to deliver 20% less to the petrol pumps.

The Council appeared to be convinced by Lubbers's argument. Den Uyl himself concluded that oil consumption must fall by 20% and petrol consumption by 30 to 35%. The Council of Ministers therefore took the decision to bring in rationing on January 7. On January 13, the general ban on Sunday motoring would lapse, and legal enforcement of the speed limit could be forgotten once rationing was in place.

However, doubts still remained over the data provided by the Ministry of Economic Affairs. By the end of November, it was clear to several ministers that The Netherlands' position was not as bad as it had been portrayed. On November 22, 1973, Van der Stoel received via Rutten a memorandum drafted by the Directorate-General for Energy, in which everything was once again worked out, with the accompanying request not to employ these figures outside the Council of Ministers. It was assumed in this memorandum that the oil companies, in anticipation of the restrictions, had increased their stocks as much as possible; and therefore, it was inferred at Economic Affairs, over ten weeks the stocks need not be considered so disastrously low.[11]

Doubts were also expressed in the Council of Ministers over the reliability of the data provided by Economic Affairs. In fact, Vredeling asked whether they depended for these figures entirely on the oil companies.[12] Transport Minister Westerterp, as he later said, was also unconvinced of the correctness of Economics Affairs' figures. 'Supplies to The Netherlands were greater than revealed by the official figures', he said. 'We had a man in the pilot service at Hoek van Holland diligently keeping a tally. At any given moment there was more oil coming in than was reported by the official figures.'[13] In spite of this, the Council of Ministers decided to switch to rationing.

The need for more accurate data concerning oil supplies did not stop here. Further information was provided in the Ministerial Council for Economic Affairs on December 5. Van Rhijn announced that the international oil companies, as had earlier become clear, were in fact more or less spreading the shortfalls over the consumer countries; the embargo was not working. The result of this was certainly not unfavourable to The Netherlands: the reduction amounted to some 20 to 25%, much less drastic figures than those presented by Lubbers in the Council of Ministers. According to Van Rhijn, only France and Britain were getting more oil. 'Switching' to benefit The Netherlands, besides, was not so simple, given the specific kinds of crude oil that refineries needed. And furthermore, not all oil companies were in a position to supply The Netherlands. Imports by Texaco were lower than other companies, while Chevron did not manage to import any oil at all.

Anyway, The Netherlands was not in such a bad situation, and yet the mood among some of those at the Ministerial Council for Economic Affairs meeting was still a sombre one. It was assumed that the embargo could still prove effective and that a general drop of some 40% could shortly be expected. Rutten, on the other hand, thought things could turn out all right. If the embargo countries did allow transit (together with all extra possibilities for *switching*), the situation would be far less serious than the gloom-mongers were assuming. Rutten was right, but the Council majority remained cautious. In fact, they remained cautious precisely because there were no accurate data to go on.[14]

Doubts over the supply data remained. During the Council of Ministers of December 10, the reliability of these figures was again raised for discussion. Den Uyl wondered if the National Bureau for Oil Products (the *Rijksbureau voor Aardolieproducten*, RBA), which was to lead the rationing, had sufficient expertise to monitor the figures from the oil companies. According to Lubbers, these data reports now fell under the operation of the Rationing Law, which made the provision of data obligatory with the backing of possible sanctions for non-compliance. The reports were now not monitored by the RBA, but the companies would be requested to do this themselves as accurately as possible. Vredeling emphasized again that the pilot service also had data at their command, although this service was not equipped for surveillance. Duisenberg added to this that customs also possessed information, but that this information could only be used in restricted circles. The Council of Ministers maintained their interest in other, alternative, figures, but the question remained unanswered as to what conclusions could be attached to such data. In any case, according to Den Uyl, what had to be prevented was the circulation of figures relating to oil stocks that were higher than those presented by the Cabinet itself.[15]

It is striking how uncertain the Council of Ministers was during these crucial weeks of November and December where the reliability of these figures over oil supplies was concerned. Time and time again there were pleas for alternative, supplementary information that could corroborate the estimates supplied by the oil companies. But apart from the need for alternative sources, the argument for using information from the pilot service seems in retrospect rather dubious, since such estimates are inevitably crude and make no distinction between supplies destined for the Dutch economy and those destined for EC partners' use. It is therefore curious that the data from customs, especially, played no part in the decision-making in ministerial council. Although there were doubts, and perhaps even a degree of scepticism, the government in any case was in no position to cut loose from its dependence on the oil companies. Still, the Cabinet was sufficiently convinced of the seriousness of the situation to introduce rationing.

Export Restrictions

Most commentaries on Cabinet policy regarding the crisis laid (and still lay) great emphasis on the domestic measures introduced to restrict the use of oil products. But that was not the only and possibly not the most effective method of combating the consequences of reduced supplies. Restricting the export from, and the transit of oil products through, The Netherlands was more effective, and was to create considerably higher percentage reductions than those achieved by the cuts at home. It involved enormous quantities of oil. In 1972, the total supply of crude oil reaching The Netherlands from abroad was approximately 130 million tons, of which circa 54% (71 million tons) was destined for further processing in The Netherlands. The other 46% (59 million tons) was immediately passed on to other countries, either by transit or re-export. Of the oil imported into The Netherlands, a considerable fraction was exported after processing: some 50 million tons in 1972.[16]

From the outset of the oil crisis, both government and other concerned bodies had been investigating possibilities of restricting, in one way or another, the transit of oil passing through Rotterdam and the export of oil products, with a view to liberating oil for domestic usage. It is not easy to get a clear picture of the reduction of transit and export in relation to the supply of crude; but the problem was discussed on several occasions in the Council of Ministers, and in other contexts too. There was the further possibility of restricting the export of natural gas, thereby cushioning the effects of the oil shortage on the Dutch economy.

On November 9, it was affirmed in the Council of Ministers that, in spite of all free market ground rules, the Dutch export of oil products throughout the EC would be subject to licensing. The way this was to operate was by applying the *Toute License Accordée* (TLA) procedure, which meant the automatic extension of licenses; but if necessary, this automatic granting of licenses could be waived. In this way, control could be exercised over the export of oil and oil products. The atmosphere in this Council was sometimes rather belligerent, for oil and gas exports were involved. Several ministers, particularly Van der Stoel and Van der Stee, proposed also bringing gas export under the TLA procedure. Others, however, were more cautious and pointed out that there was no shortage of natural gas and that such a measure would therefore be seen purely as a retaliation against the lack of European solidarity.[17]

The question of export was not only being discussed in the Dutch Council of Ministers. The same day, Lubbers and a Rotterdam municipal authority delegation discussed the possibilities of using the transit function of Rotterdam to relieve the situation in The Netherlands, for example by restricting supplies to surrounding countries, reversing the flow of the Rotterdam-Antwerp oil pipeline, and possibly supplying Rotterdam from other European ports. It was clear that such actions might well damage Rotterdam's position. If transit were to be restricted, it would have to be done very discretely.[18]

Not everyone favoured this kind of action. There was a fear at the Dutch Ministry of Foreign Affairs regarding the repercussions of any such cutback on exports. Referring to the first meeting of the Co-ordination Group, the Head of the Foreign Ministry's Department for Economic Cooperation (DES) pointed out the following in a memo to Van der Stoel. It had been made clear at that meeting that Economic Affairs wanted to apply the same export-licensing system to Belgium as to other countries. Up till then, licensing had not been applied to Benelux, even though the Belgian government had itself decided to introduce such export licenses. Since calling for consultations at the Belgian Ministry of Economic Affairs in Brussels and within Benelux had been in vain, the Dutch Ministry of Economic Affairs now wanted licensing as a countermeasure. With the assistance of the oil companies, export to Belgium could be restricted in line with domestic restrictions. It would seem a good idea, thought the Head of DES, if Foreign Affairs could 'again communicate Dutch disappointment over this turn in the course of affairs' via the Belgian embassy. There were, after all, risky aspects of 'this concealed restriction of exports', given all the Dutch effort, to maintain a common market.

The Dutch Cabinet blamed the British government for pressurising the oil companies not to divert to The Netherlands any oil bound for Great Britain. Any action in relation to Shell would therefore have to be carefully judged.

> I assume that the Ministry of Economic Affairs is in agreement, yet given the importance of the matter, it is essential in my view that the Ministry of Foreign Affairs be consulted over these aspects of the consultations with Shell.[19]

wrote the Head of DES. He also pointed out that, after some discussion the previous afternoon, the Co-ordination Group had merely agreed that a license system would have to be introduced for exports to Belgium. The reason Rutten had not then wanted to raise the possibility of voluntary restriction of exports, the head of DES supposed, was 'because it would be better if nothing could be repeated outside about this'.[20]

On November 14 the Dutch Ambassador in Brussels was instructed to request that 'high level' attention be given to this question. Van Elslande, the Belgian Foreign Minister, let it be known that he agreed with the Dutch view on this, but given the sensitivity to Arab reactions, there must be no publicity given to the matter.[21] There could be no question of any public retraction of the Belgian action already taken . The Hague did not press the matter further, but simply went ahead with the decision to bring oil exports within the Benelux trading area under license. Although the TLA principle was assumed, this move involved more than merely a formality; the licensing system opened the way to a form of export restriction that, as the Foreign Ministry had warned, was not without its consequences.

Matters did not rest at restricting exports of oil products. As we saw earlier, Lubbers had declared in the Dutch Second Chamber on November 14 that, if the EC could not function as a community where energy was concerned, The Netherlands had to defend its own interests. 'I shall involve natural gas in this', he had added. In certain situations The Netherlands must not shrink from drawing the logical inferences, or even from acting in anticipation. The Ministry of Foreign Affairs reacted coolly to this fighting talk from Lubbers.[22]

Throughout November, the export of oil products declined drastically. On November 22 a memo to Van der Stoel observed the following. Since October 23, the export of oil products, initially with the exception of Belgium, had been subjected to licensing. As a consequence, these exports had declined: for instance, the export of petrol compared with the fourth quarter of 1972 was down by some 40%, while the export to Belgium was only subjected to licensing from the beginning of December.[23] This meant an initially greater reduction than the fall in the supply of crude oil during the same period. It was also concluded at General Affairs that there had been a greater reduction in exports than was strictly justified on the basis of a proportional adjustment to the level of domestic sales.[24]

Nevertheless, at a meeting of the Economic Affairs Oil Crisis Workshop on November 21, it was assumed that in view of the drastic reduction expected in supplies, 'additional export restrictions' would have to be adopted. Before taking such a decision, the Dutch Council of Ministers had to consider fully the consequences of the reduction in domestic consumption. When it came to restricting exports, great care would be needed if the chance of international cooperation were not be put at risk.[25] In the first half of December, nonetheless, exports of oil products would drop by more than 50%.[26]

Restriction of exports applied not only to oil products but also to crude oil. Stocks of oil stored in the Botlek area were partly destined for transit or re-export. In fact, these stocks could be divided into three categories. Firstly, there was the oil for the Dutch national market and the oil to be processed in The Netherlands before being exported. Secondly, there were stocks laid up in the Botlek area under an EC agreement to maintain stockpiles on behalf of third countries and on the basis of bilateral arrangements; and thirdly, there were stocks belonging to foreign, i.e. West German, companies, stocks which in customs-technical terms were not stored on Dutch territory at all.[27]

Although no concrete measures had yet been considered, the Ministry of Economic Affairs was thinking of the possibility of somehow taking advantage of Rotterdam's transit function. In an emergency, the oil stocks of EC partners that had been stockpiled in the Botlek area under EC storage regulations could – at least in part – be requisitioned, although that would of course damage relations with the countries concerned. As we saw earlier, it had already been rationalised at Economic Affairs in October that the arrangements previously made in 1970 over these stockpiles had to be considered as merely administrative agreements of that time and were not binding on Lubbers, the current minister.

In time of need, part of the foreign stocks could thus be taken over. This was discussed in a thoroughly business-like fashion at Economic Affairs, on the assumption that the foreign emergency stocks would be sufficient to last a hundred days. As we saw earlier, the European Commission had directed member states to lay in emergency supplies for 65 days by January 1, 1971, and for 90 days by January 1974. Because it was not yet 1974, stocks for only 65 days were obligatory, and there was therefore more oil in these stockpiles than strictly necessary to meet the regulations. In the event of serious need, therefore, one could consider taking over the oil in this emergency reservoir that was surplus to strict requirements. Of course, various problems were raised by this, not least that the Dutch reputation as a transit country would be involved.[28] Nonetheless, on December 10, it was remarked in the Council of Ministers that in principle it would be possible to requisition part of the international stockpile laid up in the Botlek area.

At the end of November, the Cabinet decided not only to introduce rationing and to reduce the domestic consumption of oil products by 20%, but to look to a reduction in export of 40%.[29] A day later, it was again agreed in the Council for Economic Affairs that it was not unreasonable that The Netherlands should cut back its exports more than proportionally in order to compensate for the expected reduction in supplies. What is remarkable here is that several of those involved, as the above-mentioned Economic Affairs memorandum shows, had by now come to believe that The Netherlands was in a rather favourable position.

In a draft version of a note on energy-political action in relation to the reduced supply of crude oil to The Netherlands, which was discussed in the Council of Ministers on December 10, it was remarked that international stocks could be brought under Dutch control by government intervention; but this would seriously damage Rotterdam's trading position. And yet restricting exports was an obvious expedient. According to the text of this note, this was already occurring, and 'on the basis of the arrangement made with the oil companies'. This restriction was linked to the highly important initiative taken by the main oil companies, to try to ensure that sales were reduced equally in all countries. This unpublished passage concluded that this

> had in fact led to a greater reduction of exports than was strictly justified at this moment on the basis of proportionality with the reduction in domestic sales.

The aim of introducing export licenses had been 'to be able to control the implementation of this regulation'.[30]

As we have already seen, the predominant view within the Dutch Cabinet on December 5 was that little more could be expected of European *sharing*, and that it might even entail disadvantages. In the Ministerial Council for Economic Affairs, it was agreed that negotiating over *sharing* involved running serious risks since, without doubt, the relatively favourable energy situation of The Netherlands would be taken into consideration. A tougher method of negotiation, including natural gas exports as a counter, was therefore preferable to *sharing*. Den Uyl concurred with this view and pointed out that The Netherlands would be a very poor businessman if it were incapable of taking the political stance needed to get extra oil in exchange for natural gas.[31] These conclusions were drawn, as we saw, well before the Summit in Copenhagen in mid-December.

Where natural gas was involved, the Cabinet showed itself an assiduous promoter of Dutch interests. It was concluded in a note that a brake had to be applied to the increasing volume of natural gas exported. Where possible, that extra gas should be diverted to replace oil. The energy crisis meant that it was entirely rational to reconsider the optimal use of the Groningen natural gas. Furthermore, there had to be a clear statement of the price of gas; and of course it was obvious to assume some linkage between the price of gas and the oil prices. Lubbers was empowered to begin negotiations over a rapid conversion to this linkage. After adapting the prices, it ought to be possible to arrive at a balanced relation between oil and natural gas in the EC. If such were the case, it would then be unnecessary to bring into discussion the contracted volumes of gas exports.[32]

All in all, in the matter of the export of oil products, oil stocks and natural gas, the Dutch government had by now taken up a position that brooked few scruples. Although Heath, the British Prime Minister, had been reproached with his tough and purely self-interested attitude, the attitude of the Den Uyl Cabinet was essentially no different. Initially, it had still been hoped that there might be mutual cooperation within the OECD or the EC. Once that hope had evaporated, the Dutch government's positions hardened. Given the relatively favourable situation The Netherlands found itself in, with the American offer of support a very welcome boost, however illusory it might be, the Dutch Cabinet seemed to have lost all interest in *sharing*. Further, it had for some time no longer been considered in the nation's interest.

Preparations for Rationing

In fact, it is all the more remarkable that, despite this relatively favourable position, the Dutch Cabinet was the only West European government to switch to the introduction of rationing. The initiative for this move came from the Ministry of Economic Affairs. As early as November 7, the Oil Crisis Work Group came to the conclusion that the best way to cope with the problems that had arisen would be to introduce a rationing system as rapidly as possible. It seemed to the Work Group that an extension of the ban on motoring would affect road transport too adversely, with all the ramifications that would have.[33] This assessment was adopted by the Co-ordination Group the following day; and one day later the Council of Ministers also decided to set in train the preparations for rationing.[34]

There were objections expressed, notably by the Ministry of Justice, but from that moment on, the introduction of rationing was pursued indefatigably. On November 13 an important step was taken by activating the National Bureau for Oil Products (the *Rijksbureau voor Aardolieproducten*, or RBA), which according to the stipulations of the Rationing Law would direct the rationing of natural gas products. Initially, the RBA was preoccupied with the difficulties and the complaints caused by the reduced supply of oil products. This was often dealt with by officials temporarily detached to the Bureau. At a later stage, the RBA played a central administrative and informational role in preparing for the implementation of rationing.

On November 16, however, the Council of Ministers hesitated over whether to continue with the introduction of rationing at short notice, even though Economic Affairs argued strenuously that if it did not happen before December 17, because of technical problems, the entire operation would have to be postponed till January. But the Council was more cautious than Economic Affairs. At that moment the oil supply was unchanged: there was no question of eating into reserve stocks.[35]

The oil companies, however, thought it inevitable that effective restrictions would have to be introduced quite shortly. Pressure was thus being exerted on the Cabinet. It was also argued in a memo from one of the council advisors from General Affairs that fundamental decisions had to be taken before the end of the year; it was not possible to wait any longer.[36] But the Council of Ministers decided that the introduction of rationing was not yet possible because of the multiple legal complications and transport-technical problems. There would first have to be consultations involving, in any case, the Ministries of Justice, Finance, and Transport & Water Management. And so it turned out.

Five days later, the Co-ordination Group again pressed the case for a swift introduction of petrol rationing. A number of decisions over data would have to be taken within a very short time, as December 10 was the deadline for beginning to distribute coupons. This meant that authorisation forms and circulars would have to be sent to local authorities as soon as possible, particularly in connection with supplementary allocations. In the light of the deteriorating situation, the Co-ordination Group thought any postponement of preparations at this stage would be irresponsible. Supplies were now beginning to decline, and the forecasts were gloomier than ever, or so it was assumed. The oil companies also found further restrictions on consumption necessary. Postponement would mean that introduction could not begin until mid-January.[37]

And yet on November 23, the Council of Ministers once again decided to wait, even though preparations had gone ahead and approval had been given for the Ministry of Transport and Water Management to send out all the authorisation forms. The Ministerial Committee for Oil Problems was also instructed to prepare the rationing as effectively as possible. The first discussion over this took place on November 29. The idea was to meet every Thursday to discuss, among other topics, the reports issued by the Co-ordination Group, which met every Wednesday afternoon. But the definitive decision to implement rationing was postponed till November 30. There were various problems to sort out in the meantime, measures that would have to be prepared in advance, including the car-free Sunday, public transport, freight transport, the costs of rationing and the bearing of the anti-discrimination regulations of the EC on the question of foreign motorists.[38]

At the end of November, the Oil Crisis Workgroup at the Ministry of Economic Affairs concluded that petrol rationing would have to reduce private motoring by about 40% and the use of cars for business purposes by some 20 to 25%, yielding an overall reduction of some 30%.[39] These objectives were accepted by the Co-ordination Group and by the Council of Ministers. As a consequence, on November 30 the decision was taken by the Council of Ministers to introduce petrol ration coupons with effect from January 4, 1974, and also for LPG and propane. A reduction of 25% was thought appropriate for the latter two types of fuel, for petrol 30%. Diesel was not included in the rationing scheme on account of the importance of freight and public transport, though controlled sales of diesel oil were thought necessary, to be achieved by the main companies limiting their deliveries.

In the first instance, it was thought that rationing would be introduced for a period of four weeks. Because of all the preparatory work involved, this could not be introduced in December, the original preference of Economic Affairs, but would have to wait until January. Meanwhile, Lubbers could distribute to the oil companies all the necessary guidelines concerning petrol deliveries that would lead to a 20% reduction at the petrol pumps. Various other measures were also prepared to anticipate the consequences of rationing, in particular to guarantee the maintenance of public transport and to stagger working hours.[40]

These measures prepared by the Cabinet were set down in a note for parliament. In *Restriction of the oil supply and its consequences*, the packet of measures put together by the government was set out in a list, on the basis of an assumed 50% reduction in the oil supply by January, followed by a further 30 to 35% drop in the following months. This note was based to an important extent on the material supplied by the Central Economic Committee. The measures themselves, both short term and longer term, were as follows:

- limiting petrol consumption by some 30% through rationing;
- substituting oil consumption at Hoogovens and electricity-generating stations by coal, and by extra use of natural gas;
- switching to natural gas for domestic use, in horticulture, and large-scale users;
- accelerating programmes of exploration and exploitation of natural gas and oil;
- encouraging the insulation of homes;
- lower heating levels, and reducing electricity use for lighting by 10%.[41]

Over the course of December, various other decisions were taken. A government decree restricting the use of electricity for commercial lighting was issued, with the result that over the Christmas period the major cities, as elsewhere in Europe, assumed an atmosphere reminiscent of the 1950s as the illuminated advertisements were switched off. Because diesel did not fall under rationing, the Ministry of Transport & Water Management called on freight vehicles to observe a speed limit of 80 kms per hour, under the slogan *'Plankgas oliedom'* ('Foot down, fuel-dumb').

But there were still knots to cut through. It was announced that rationing applied to delivery vans and small trucks and lorries. Passenger cars with a diesel engine escaped the rationing, but most mopeds did not. Furthermore, it was decided that all passenger cars, irrespective of engine capacity or weight class, would be allocated fifteen litres per week. This basic assumption, the same petrol allocation for all classes, ran counter to the rationing plans in the filing cabinets of the Ministry of Economic Affairs. The Secretary of the Co-ordination Group, Willemsen, later gave as his opinion that this kind of equal treatment for everyone was a typical expression of the Den Uyl Cabinet's 'sharing philosophy'.[42]

Meanwhile, in mid-December a working group of officials drawn from the Ministries of Economic Affairs, Foreign Affairs and Transport & Water Management drafted a regulation for foreign motorists. To the satisfaction of the Foreign Ministry, the European Commission was informed that The Netherlands had adopted a liberal regulation for foreigners that involved no hold-up at border crossings. The thinking behind this was the anticipation that reciprocity would be observed should petrol rationing be introduced in other EC countries. This argument for a non-discriminatory treatment of foreigners was repeated in the broader context of the EC Working Group for Transport and Energy. In addition, it was agreed there to extend preference to international motorway and air transport, as well as to domestic shipping.[43]

The Enabling Act

In the meantime, the Den Uyl Cabinet was busy getting the heavy artillery in place to deal with the consequences of the embargo. On December 8 the Cabinet submitted to the Second Chamber the proposed legislation, the *Enabling Act*, or to give its full title, the *Act enabling legislation to regulate the creation of income and protection of employment in 1974*. The aim of the Act was to allow intervention in wages and salaries, and in conditions of employment, as well as prices. In the Explanatory Statement accompanying the Act, it was explained that the oil crisis, and the economic problems arising from it, such as growing unemployment, had necessitated the introduction of the Enabling Act.[44]

It was pointed out earlier that the Enabling Act cannot be seen purely as the consequence of the oil crisis, even though the Cabinet continued to take for granted the connection between the Act and the specific measures designed to deal with the oil crisis.[45] The law came into force in a period of tense socio-economic relations and of inflationary tendencies. In the previous year, the Biesheuvel government had striven – with limited success – to establish a general agreement between employers' and workers' organisations in an attempt to gain control over rising prices and incomes. Attempts at wage restraint, however, had been rejected by the trade unions.

In the spring of 1973, differences of opinion over index-linkage led to serious conflict between industrial unions and employers. Despite strike actions, the employers did not yield, leading the Union Federations, the NKV and NVV to withdraw from talks in the Social-Economic Council and the Joint Industrial Labour Council. In the autumn of that year, because of the oil crisis, negotiations over a general agreement for 1974 took place under rather dramatic circumstances. These negotiations led to new and serious disagreement over complete index-linkage. Under Cabinet pressure, a compromise was reached on November 5, but this was rejected the following day by the rank and file of employees' organisations. Consequently, the unions announced that any further central discussions were pointless. Promises earlier conceded by the employers' organisations, it was insisted, would have to be adhered to at the collective labour agreement negotiations, which would now have to take place separately for each branch of trade and industry.

The Dutch Minister of Social Affairs, Boersma, also took the position that the accord of November 5 would still have to be implemented. The Minister told the Chamber that the government was of a mind to bring in an Enabling Act, on the basis of which binding conditions of employment could be laid down. The November accord, he said, would be taken as the starting point for government action. The government submitted its Bill to the Second Chamber on December 8. It was explained that a number of powers had been included in the Bill in order to act with the necessary decisiveness needed to bring in adjustments to the social and economic order, specifically in those areas where existing powers were either inadequate or non-existent. The areas in question were pay and other conditions of employment, foreign labour, the conditions of employment of civil servants, the salaries in the professions, dividends, leases and rents. Furthermore, the Enabling Act embraced a number of changes in existing laws, most notably in the Price Law.[46] The most important reason for submitting the Bill to parliament was the oil crisis, which threatened the provision of energy and the employment situation, and the steep rise in the price of raw materials, leading to fears of inflation.

The Enabling Act was speedily dealt with by parliament, though certainly not without occasionally heated discussion. Moreover, both Chambers added their different amendments to the Bill, which before being submitted had already been modified on advice from the Council of State. The Bill was discussed by special committees within a matter of days; and in mid-December followed the memorandum in reply and the first government amendment, to be followed several days later by three further amendments.

The PvdA and to a lesser extent the ARP supported the Bill. In contrast, the VVD was not entirely convinced of the need for the government to be given such a mandate to intervene in socio-economic relations. According to the VVD, the oil crisis had been seized upon as a means of pushing through a social democratic pay policy. Furthermore, in the view of the VVD, parliament had been by-passed, a point of view also shared by the Christian Democratic parties. In response to this criticism, Den Uyl assured parliament that the intention was in no way to achieve specific incomes-political objectives. It was a matter of an emergency that had been created by the oil crisis. Although this explanation was accepted by the majority, a series of amendments were introduced by the Chamber. In the first place, it was arranged that Boersma must confer with the Ministry of Economic Affairs in different situations. Evidently, the Chamber wanted to defend the position of Lubbers and Economic Affairs. Furthermore, the Social Economic Council and other consultative organs must be involved in some decisions. And finally, parliamentary control should be strengthened, which meant, among other things, insisting that certain government decisions must be reported to the Chamber.[47] On December 20, the Second Chamber voted by a majority to pass the Enabling Act. The First Chamber, however, decided to deal with the Bill only on January 8 and 9.

Meanwhile, the Cabinet had sent the Second Chamber a second note over the oil crisis, entitled *Policy note limiting consequences of oil shortage.*[48] This note, also partly based on material delivered by the Central Economic Committee, was comprehensively dealt with in the Ministerial Council for Economic Affairs. The aim of the note was to flesh out the intentions announced in the first note concerning the oil shortage and in the Enabling Act. The Cabinet reiterated that, in view of the great uncertainty over oil supplies, it was inevitable that measures be taken to control prices and wages and the growth of government expenditure. The restricted oil supply and the rise in oil prices would contribute to a steep rise in prices and wages, coupled with stagnating or falling production, productivity and imports/exports. Apart from the measures announced earlier, the note announced capital spending on the extra commitment to natural gas and on increasing the profits from natural gas. With regard to industrial relations, the Cabinet announced measures aimed at maintaining the level of employment as far as possible, such as support for businesses finding themselves in difficulties, and the creation of new jobs, specifically a programme of insulation in the building industry. Rising wages must be carefully monitored (even though wages in 1974 would increase considerably), possibly requiring the introduction of a 'rest period' in connection with the uncertainty surrounding oil supplies.

Doubt and Postponement

At the end of December, another wave of doubt assailed the government over whether the rationing, already prepared and at the ready, was really necessary. The forecasts concerning oil supplies were more optimistic, even within the Council of Ministers. On December 21, for the last time in 1973, the question of the oil shortage was raised in the Council of Ministers, and with it new data on current supplies were provided. During the handling of the Enabling Act the previous day, with the presentation of information on the oil situation, the Cabinet had 'passed through the eye of the needle', as Den Uyl remarked.[49] Den Uyl pointed out in the Council of Ministers that there was a general distrust of the figures. Could not more exact figures be obtained from the Customs and pilots services on the New Waterway, so that the Cabinet might be properly informed of the situation?

It was clear in this ministerial council that Lubbers was no longer convinced of the need for rationing. In the following weeks he would more and more explicitly wonder whether it would be more sensible simply to blow the whistle on the entire operation. During these weeks it was predominantly the PvdA ministers, not least Den Uyl himself, who would persevere with putting rationing into operation in spite of Lubbers's reservations.

Lubbers pointed out that there had been a one million tons deficiency in December's incoming oil supply, which was much better than the earlier prognoses. His impression was that the oil supply in December had been better than was expected at the end of November. The situation was also better because road transport had been effectively cut back, the oil companies commanded vast stockpiles, and more oil than anticipated was still arriving. He concluded that it would be better to postpone rationing, even though from the point of view of policy continuity it might perhaps be better to carry on with it. If the rumours of huge oil stocks were correct, the Cabinet would have to reconsider its policy.[50] Once again it was evident that the Cabinet was not in a position to command an adequate picture of the situation; and nowhere was this more in evidence than in Lubbers's own doubts as to whether rumours of overflowing oil tanks in the Rijnmond area were true or false.

These doubts grew. In early January, NRC *Handelsblad* started questioning the government's efforts to introduce rationing, predicting a tide of protest from trade and industry. There was also doubt in the Second Chamber, the paper thought, over the necessity of rationing. Very little inconvenience had so far been suffered as a result of the embargo and reduction in oil supplies. Lubbers himself had said that he was not pessimistic about this. The oil supply was, in his view, still reasonably up to the mark.

> Clearly, from this statement, the government must now realise that the situation with regard to the oil supply is better than anticipated. Why then persist with a complicated and, for many, inequitable system of rationing?[51]

Such doubts as to the need for rationing spread within the Council of Ministers at the beginning of 1974, doubts mainly fed by Lubbers. There were technical problems, he said, such as the allocation of petrol: such allocation could only be completed two weeks after rationing had been introduced, causing great uncertainty for many applicants. But the most important reason for postponing or even cancelling rationing was for Lubbers the fact that oil stocks on December 31 were 'virtually intact'. Supply, moreover, was greater than anticipated. At any event, he proposed postponing rationing by three weeks.

This went too far for the majority of the Cabinet. The Council of Ministers decided to compromise with a one week postponement, with no shifting of this deadline. The consequence of this decision was important and remarkable, for it meant that the first period of rationing would last only three weeks, not four. The total quantity of petrol that could be used in this period, however, remained the same as originally provided for four weeks, which in turn implied that every private vehicle now had the right to twenty rather than fifteen litres per week.[52]

It was a decision that did the government's reputation no good. The press accused the Cabinet of irresolution and panic; 'Poorly organised', ran the headline of the lead editorial in *de Volkskrant*. 'What is hard to swallow is the government's panicky play from one day to the next.' Ministers contradicted one another; the measures taken were deficient, and many administrative problems not foreseen. The petrol ration suddenly went from sixty litres per four weeks to sixty litres per three weeks.

> Analysing this policy – if policy it can be called – leads one to conclude that the government set sail on its mission with nothing better than a wet finger held up to the wind.[53]

'Unfortunate' was how *Trouw* characterised the five-day postponement of petrol rationing: not only the time period, but part of the reasoning also seemed at fault. 'Why did the government have to wait for this postponement until two days before rationing should have been implemented?' Public transport could no longer reverse the extra schedules. Inevitably, and justifiably, there was huge irritation. Furthermore, the reason – that the oil supply had slightly improved – did not hold water.

> The Cabinet has always acted as though it was incapable of tuning its rationing policy to such minor fluctuations... By giving the impression now that it can, the Cabinet has provided with ammunition for some time to come to its many critics who will soon be wanting to see daily proof of that rationing is necessary.[54]

NRC *Handelsblad* was also critical. 'The unexpected U-turn in policy does not suggest well-considered policy-making,' was the paper's judgement. The question of whether rationing was in fact urgently needed had not been adequately answered by the Cabinet.[55]

The Cabinet Ploughs on

On January 11, the decision was taken in the Council of Ministers to press ahead with the prepared rationing scheme, in the first place for three weeks, from January 12 till February 3. It was by now believed that this rationing would lead to a saving of 20%, considerably less than the 30% previously assumed. If this 20% saving was no longer necessary, the system would revert to one or two car-free Sundays in combination with a speed limit of 90 kms per hour.[56]

In the Council of Ministers of January 11, confusion once again reigned over the true figures for stocks and supplies. Lubbers informed the Council that, in any case, supplies at the beginning of the year were greater than anticipated. Consumption now lay in the region of 67% of its normal magnitude, while supplies amounted to some 70%. Lubbers's main concern was the development of prices, not oil supplies. Den Uyl, however, disagreed. In his view, the December supply had been lower than had been reported at the time, in fact some 50% lower than the previous year. In public, however, a figure of 30% had been discussed in order to avoid too much disquiet.[57]

Den Uyl's line of reasoning contradicted the earlier statement from Lubbers, that oil stocks were more or less intact at the turn of the year. Furthermore, from Lubbers's report it could be construed that domestic oil use had been reduced by roughly the same amount as the oil supply; and finally there seemed to be a clear improvement in the provision of oil. And yet the decision had been taken to implement rationing, even though the Council of Ministers accepted that it need not last long.[58]

The Cabinet decision seems, certainly in retrospect, highly contestable. Stocks were being maintained, the oil supply was recovering, and there was already on the cards a considerable reduction of exports and reduced consumption. The circumstances in which the Cabinet decided to go ahead with petrol, LPG and propane rationing are perhaps best illustrated by a telex that same day by BK Gas to the National Bureau for Oil Products (the RBA), asking that the rationing of LPG be postponed until genuine shortage could be detected. According to BK Gas, stocks for the coming months looked good. In fact, given the excess stock, there was the danger that a good deal of unsold gas would 'have to be burnt'.[59]

A last striking aspect of the introduction of rationing was that, at this point, the government clearly began to deviate from the policy implemented by surrounding countries. The announcement of rationing had caused surprise in the West German Republic, as the Dutch Ambassador in Bonn reported to The Hague; for there it was considered that the car-free Sundays and the speed limit would be sufficient. Nor was rationing introduced in Belgium. The only other European country that had introduced rationing was Romania.[60] This fact that countries bordering The Netherlands had not found it necessary to introduce rationing would soon prove an insuperable problem for the rationing experiment.

Why was rationing implemented, in spite of all the ambivalence and in spite of the increasingly optimistic data? By now, Lubbers was no longer advocating it and was essentially expressing doubt. According to Rutten, this was mainly attributable to Lubbers's uncertainty.[61] But on the other hand, there were also good reasons to doubt the sense of rationing. It was mainly the PvdA ministers, Den Uyl above all, who pushed through the decision. However, beside this, one has to set the role of the Co-ordination Group and of Rutten himself, as chairperson of the Group. Rutten had also argued for implementation at the gathering of the Ministerial Committee where he had represented Economic Affairs in the absence of Minister Lubbers.[62]

The majority of the Cabinet, for a variety of reasons, did not share Lubbers's view. One argument was the uncertainty of supplies in the near future, despite the satisfactory figures on parade. A second argument, advanced by Van der Stoel among others, was the political instability in the Middle East, which might in time lead to the announcement of further restrictive measures. A third argument, also stressed by Van der Stoel, was the fear of repercussions from the Arab side should they perceive from the Cabinet's decision-making that the embargo was ineffectual.[63]

And yet these considerations were not the main reasons for pushing ahead with rationing. The most important point was the credibility of the Cabinet's policy. It was concluded by several members of the Cabinet, not least Den Uyl, that postponing rationing, and especially any ostentatious change of policy, would create a very bad impression with the Dutch electorate and arouse doubts as to the government's strength of purpose.[64] This, however, certainly after the postponement of the previous week and the public reactions to that, was a rather contentious argument.

The idea that rationing was mainly an exercise of a social democratic, centrally controlled, economic policy of austerity, which was for example the view subsequently expressed by the then director of Shell-Netherlands, Helfrich, is surely an exaggeration.[65] Rationing, after all, can hardly be seen as an example of progressive economic politics. The introduction of the Enabling Act, passed by the First Chamber in the second week of January, went much further in that direction. Although the introduction of rationing was championed in Cabinet by the PvdA ministers above all, it was the bureaucratic weight of all those preparatory measures that played such an important role in the decision to keep on going. Even Secretary-General Rutten of Economic Affairs was in favour of pressing on; there was simply too much civil service labour invested in the preparation.

Three Weeks Rationing

The rationing system was in fact based on a vast mass of paperwork. In the final report of the National Bureau for Oil Products (RBA), it was later found that there had been 47 different forms in circulation, two types of permit, seven kinds of ration card, one authorization, five categories of ration coupon and three kinds of allocation. Furthermore, a large number of government and other organisations were involved in the whole operation: various ministries, municipal and provincial institutions, the

National Traffic Licensing Authority, the Central Office for Motor Vehicle Taxation, Chambers of Commerce, the Dutch motorists association ANWB, not to speak of all the petrol pump owners and oil companies.

In the allocation of coupons for the initial rationing period – originally four weeks – it was assumed that these would remain valid for six months. The disadvantage of this long-lasting validity was the increased risk of fraud, but that was simply accepted as part of the costs. The great advantage was the flexibility afforded, and for this reason trade in coupons was not forbidden. Allocation was determined on the basis of data supplied by the Office for Motor Vehicle Taxation. A total of 3.1 million ration cards were issued. The RBA made full use of the possibility of involving other governmental bodies on the basis of their expertise, specifically in issuing extra coupons for a whole series of exceptional categories: business vehicles, commuter traffic, rental vehicles, invalids (who would need to produce evidence of their invalid status). Where road traffic was concerned, this authority was the Directorate-General for Road Transport (under the Ministry of Transport and Water Management); for the agricultural sector it was the Directorate-General for Food and Agriculture; and for the diplomatic service, the Cabinet and Protocol Department in the Foreign Ministry. The apparatus of the Directorate-General for Road Transport quickly proved inadequate to the task. Out of necessity, 93 distribution circuits subsequently had to be set up to issue supplementary allocations for the business use of private vehicles. Inevitably, there was confusion among the public as to which was the appropriate authority to turn to. There was also considerable uncertainty over what exactly qualified for extra allowances. By shifting the date of implementation to January 12 and because of the simultaneous improvement in the general outlook, there was also more room to improve the allowance of extra petrol.[66]

Soon enough, however, rationing proved an efficient means of generating political and social discontent. As early as January 8, the Cabinet policy came under fire in the First Chamber: the need for the entire rationing operation seemed so dubious. The Cabinet stood its ground although Lubbers declared that a reduction of 20% would be sufficient rather than the 30% deemed necessary in the December note. A motion was submitted to change government thinking but was rejected by 45 votes to 24, despite the support it received from eight representatives of the ruling parties.[67]

Against this background, the debate in the Council of Ministers over January 18 and 19 centred on the question of whether rationing should last beyond three weeks. There were immediately differences of opinion evident. Westerterp cast strong doubt on the need to continue. Business traffic, in his view, was encountering considerable hindrance, while the agreement between Egypt and Israel over disengagement gave cause for optimism. Lubbers seemed to be in doubt. Den Uyl, in contrast, thought that as long as the oil supply was less than 80% of the previous year, it was necessary to prolong rationing by a further three-week period. This, in the end, was what was decided.

In practice, however, rationing rapidly lost its effectiveness. Along the Dutch borders, motorists were crossing the frontier to buy up foreign petrol on a large scale, resulting in a drastic slump in sales at the pumps in The Netherlands. As a result, the Dutch petrol stations were reverting to the sale of petrol without coupons. Nor was this flouting of rationing restricted to the pumps in the border areas. In increasingly wider circles, it was felt that the rationing system was being discredited. At this stage, during talks between the National Bureau for Oil Products and the oil companies, it appeared that the difference between supply and demand was estimated at 15%, less than the 20% limit set by the Cabinet.[68] The question was therefore what purpose the colossal effort of paperwork actually served. On January 22 the General-Secretary of the Cycle and Motor Manufacturing Industry (RAI), W. Hustinx, pointed out to Lubbers in a telex that calculations showed that, on an annual basis, 4866 million litres of petrol had been distributed under rationing, not including extra allocations for police, fire service, invalids etc. The total petrol consumption for 1973, had there been no oil crisis, would have been 4865 million litres. In short, rationing was pointless. The RAI therefore would prefer, as Hustinx delicately remarked, that the dislocation of the motor industry and trade could be ended by reverting to more normal relations, on the basis of voluntary savings in petrol consumption.[69]

In the second week of rationing, it became clear that the organisations involved, the oil companies as well as the distributors and pump owners, were simply not sticking to the prescribed rules. On January 22, the Association of Automobile Dealers and Garage Owners told Lubbers in no uncertain terms that they were no longer willing to cooperate in the 'rationing circus'. Many petrol pumps had for some time gone their own way; some had even succeeded in importing petrol from abroad. But the companies were also delivering generously. On the basis of a current account system, the oil companies were allowed to supply the pump owners with extra petrol, which would be taken into account subsequently on return of the coupons from sales. This regulation was probably rather freely interpreted by some oil companies. It was subsequently concluded by the government that the current account system had probably let 'too much air' into the rationing system.[70]

On January 21, Rutten was talking of 'rebellion'. The question was how the government would react. The view within the Ministry of Justice was that it would not help to maintain rationing with a heavy hand. The legal possibilities inherent in the situation were complex. Furthermore, there was a good chance that rationing would be abolished within days, which would make the Ministry of Justice look a little ridiculous. Discontinuing rationing completely, however, seemed equally undesirable. After all, the Cabinet had just announced three further weeks of rationing. It was thought at the Ministry of Justice that the Foreign Ministry was also against abolishing the scheme as long as the embargo against The Netherlands was not lifted. A system of extra measures and exceptions was thought too complicated. The best way out therefore seemed to be 'a temporary suspension of the disposition'.[71]

Meanwhile, the estimates of stocks and supplies were looking even more favourable. On January 9 it was still being assumed by the Co-ordination Group that supplies in January and February would be 30% down on the previous year. But two weeks later, the shortfall was assumed within the Ministry for Economic Affairs to amount to only 22%. It was expected on January 9 that 28% less crude oil would be processed in The Netherlands; whereas two weeks later the figure cited for this reduction was 19%. In fact, it was concluded on January 24 that the outlook for most oil products was favourable.[72] On January 17 the Ministerial Committee therefore decided to go over to publication of the figures, though there was some doubt on this score within the Foreign Ministry. The Head of the Department for Economic Cooperation (DES) pointed out to Van der Stoel that there remained some doubt about the assumption that the Arab countries by now no longer had any objection to the redirection of the oil stream. This idea had played a part in the decision of January 17 but it was by no means certain that a government demonstration that the embargo had been ineffectual would not irritate some Arab countries. On the other hand, the Head of DES appreciated that there were good domestic reasons for resuming publication of the data on oil supplies.[73]

On January 23 the situation in the Council of Ministers had changed from four days earlier, when it had been decided to extend rationing to February 25. A memo from Rutten to Lubbers was submitted in which it was confirmed that 'rationing was being circumvented in diverse ways and on a vast scale'. Meanwhile, Van Agt concluded in a letter to Den Uyl 'that maintenance of the ban on supplying petrol other than on production of coupons had gone completely haywire'. There were far too many coupons in circulation, which was mainly the consequence of their lavish distribution for road vehicles. There was unrestrained sale of petrol in border areas, given the competition from free trade in fuel over the frontiers. Van Agt noted that Den Uyl had apparently proposed a ban in order to import more extra petrol from abroad; but the law did not provide for this, quite apart from the problems with the other Benelux countries and the EC.[74]

Den Uyl concluded that the pressure from both parliament and the country to dispense with rationing at once was growing because it was being assumed that there were sufficient stocks of petrol. This assumption was fostered by remarks by the head of the Rotterdam trade unions, A. van Schravenmade, who had stated that the reserve tanks in the Botlek Rijnmond area were more than full. The oil companies had let it be known that a saving of 15% would be necessary in the coming weeks, less than the 20% that had served as reason for the Cabinet to introduce rationing. The Co-ordination Group had meanwhile proposed, in line with the Cabinet decision of the previous week, to call a halt to rationing on February 25; but Den Uyl was now considering a much earlier date, viz. February 4. Pronk put forward the most explicit objection to this: there were no reasons for an immediate cessation. The supplies, after all, had not become much better.

To most members of the government, however, it seemed the situation was becoming uncontrollable. Van Agt again emphasized that large numbers of cars were crossing the borders to fill up in neighbouring countries. The result was more and more petrol pumps were supplying petrol without coupons. Maintaining rationing until February 24, in his view, would be impossible. He shared Den Uyl's point of view that the scheme should be wound up on February 4. Nonetheless, there remained doubts to the last. There were in fact two draft decisions on the table: one to extend rationing by a further three weeks, the other to stop. In the end, the council decision was to end rationing on February 4, but to continue with efforts to reduce petrol consumption, specifically through the maximum speed limit.[75]

The series of decisions taken by the Den Uyl Cabinet ensured the inevitable outcry in the press. *De Volkskrant* had many questions it wanted answered over the ending of rationing. Why had the Cabinet found it necessary only one week previously to decree a new period of rationing? Had the oil supply unexpectedly become so much more favourable? With due respect to all conceivable arguments, the paper retained the impression 'that the early abrogation of rationing is rather to be attributed to the growing chaos than to a better than anticipated oil supply'.[76]

The paper expressed itself even more forcefully two days later. After the announced end of petrol rationing on February 4, the remaining period under rationing had become a joke. *De Volkskrant* spoke of 'government drift': certainly, the paper said, the epithet 'bungling' seemed well-deserved for the policy of the recent period.[77]

The story of rationing seemed to the eyes of NRC *Handelsblad* both brief and strange. Lubbers had not been able to make completely clear why it had been decided to abandon rationing. In spite of the encouraging figures, it hardly chimed with the long-term objectives that Lubbers had always presented.

> It is not so long ago that optimistic views of the oil supply in *The Economist* were being rubbished by various members of the Cabinet.[78]

Two days later the paper returned to this question. The cancellation of rationing had been more or less dictated by the oil companies and the Association of Automobile Dealers and Garage Owners (BOVAG), with the companies playing a particularly significant role. They had initially promised the government to cooperate with implementing instructions through to the end of the rationing period. In fact, because of the extraordinarily flexible delivery rules for petrol pumps, implementation of rationing had been less than scrupulous. Pump owners had in various ways managed to lay in extra quantities of fuel. Furthermore, as it turned out in practice, the stipulation that petrol could only be supplied on the production of coupons had generally been ignored. The whole affair only served to illustrate the shortage of trust between companies and government.

> This is perhaps explained by the growing belief that the oil companies have provided government with insufficient insight into the circumstances and the consequences of the oil crisis.[79]

Once it became clear that rationing would be abandoned, it became even more difficult to maintain any sort of control over the petrol pump owners, as was clearly realised at the Ministry of Justice. In a memorandum to the Secretary-General at the Ministry, it was suggested that it would be inconceivable to come down hard on the pump owners. But the real villains had undoubtedly been the Association of Automobile Dealers and

Garage Owners (BOVAG), and a prosecution of this organisation might well help remove the impression of impotence. The Cabinet could demonstrate that important issues were at stake if a monopoly organisation like BOVAG resorted to 'civil disobedience'. But of course there were also risks attached to such a judicial course of action. The most important policy objection was that by involving the courts, government would be manifesting a rather convulsive need to justify itself in what was, after all, a fairly feeble matter.[80]

This, then, was how rationing ended: in civil disobedience and chaos. There was much discussion at the Ministry of Economic Affairs over the question of whether it needed winding up and the cost counted; but in the end the instructions were simply to leave matters as they were. The Coordination Group also later admitted that there had been far too many problems involved. The group thought the way the oil companies had acted had been very variable: Shell had probably exploited the opportunity to restock (as was later established in the Council of Ministers). The Coordination Group concluded that petrol consumption in January had been approximately 20% less than normal.[81] The target figure had thus been achieved, but the question was whether this might not have been achieved even without rationing. Furthermore, it turned out that the combination of this reduction in consumption together with the recovery in the oil supply was by now causing storage problems.

Following the end of rationing on February 4, there were initially no additional measures prepared. It was a matter of re-introducing the car-free Sundays, at least once or twice a month, and the legal fixing of a maximum speed limit. Even these measures quickly became superfluous, however, for there was now no shortage of oil. When, at the end of January, the possibility of improving both oil provisions and The Hague's image was considered at Economic Affairs, the question was raised whether there was actually any need for an increase in supply. The answer was 'no'.

> In view of the unusually mild winter, the ban on reporting incoming tankers, information that tankers are being diverted to other ports, as well as the rumour that old tankers are being filled offshore because of lack of storage space on land, this question can be answered in the negative.[82]

Helfrich later confirmed the truth of these rumours that old tankers had been used to store oil on the North Sea. Furthermore, Shell had instructed tankers *en route* to Rotterdam to sail slowly, since their oil could not be stored in the Botlek area.[83]

Conclusion

The storage problem stands in stark contrast to the apprehensive suspicions voiced in the double Cabinet note to parliament in mid-December. For this reason alone, according to various authorities involved, it was not surprising that rationing had been a failure. According to one note from the Ministry of Justice, during the three weeks of rationing petrol had been widely sold without coupons and without any corrective action being taken. Not that it would have been easy to take effective action, given the economic unity within the EC and free traffic between Benelux countries. More generally, the note argued, a rationing system can only work in times of scarcity, and during these three weeks there had been no genuine scarcity.[84]

More radical criticism was expressed in a report on the experience of the Amsterdam distribution circuit. The government had hesitated far too long, and the procedures by which rationing was implemented had been inadequate and confusing. Above all, the decision to reduce the rationing period from four to three weeks, while the allowance itself remained unchanged, had seriously undermined the credibility of the whole rationing scheme. Different members of the government, moreover, seemed regularly to contradict one another, which had made a very poor impression on the public.[85]

The National Bureau for Oil Products (the RBA) subsequently concluded that rationing was a means of defending scarce basic materials as effectively as possible under conditions of war and concomitant major shortages. But because such problems no longer existed by January 1974, the public had no belief in the legitimacy of the measures. Petrol was sold illegally on a wide scale, and the government turned out to be in no position to take any countermeasures. More generally, serious mistakes were made: information had been inadequate, co-ordination had been lacking, there were far too many coupons in circulation. In fact, a great deal had gone wrong.[86]

But the most important problem was the fact that stocks and supplies in January were in so much better health than initially assumed. Why did the situation turn out so favourably? Here it is relevant to pay attention not solely to the figures for supply, for in December these were indeed scarcely encouraging. In that month, supply fell back by a huge percentage in comparison with December 1972. And yet at the beginning of January, Lubbers was able to establish that stocks were virtually intact.[87] There could have been several reasons for this striking fact. In the first place, it was evident that some companies, with an eye to the coming winter and because of an anticipated rise in oil prices, had transported extra stocks of crude oil to The Netherlands. The RBA even spoke of 'record stocks'.[88] In the second place, the export of processed oil products was in all probability drastically reduced. Thirdly, there was a drop in the domestic use of processed oil products as the result of restrictive measures, and on the basis of voluntary action, also helped by such coincidental factors as the mildness of the winter. Fourthly, supplies recovered in January much faster than expected as a result of the *switching* policy operated by the oil companies. Furthermore, it became clear that the embargo was not everywhere being effectively enforced. And finally, after the price rises of December, the production limits were made more flexible by the OAPEC countries.

By the time it became apparent in January that supplies were beginning to recover, rationing had in fact become superfluous. Lubbers, who had originally been a champion of the rapid implementation of rationing, had by then come round to the position that the entire operation was too late in the day. And of course he was right. The system quickly collapsed when several of the players involved simply chose not to abide by the rules: the Cabinet was confronted with 'civil disobedience' on a wide scale. Even the oil companies – or at least some of them – seem not to have stuck to the distribution regulations, and as a result the major companies played a remarkable role, since they had originally argued for more radical measures than the car-free Sunday and voluntary speed limits. Even the Cabinet, in justifying the introduction of rationing, later referred to 'the darker prospects outlined by the oil companies at the time of the decision-making in question'.[89] And in all probability they continued to paint this dark picture until well into December. When the government decided to put into operation the rationing mechanism prepared so long in advance, the oil companies appeared by then to be no longer convinced of the need for any drastic reduction in consumption; in fact, they contributed to the evasion, discrediting and abandonment of the whole rationing scheme.

The Den Uyl government was always one step behind events. This was partly the consequence of the fact that the Cabinet was so dependent on the information provided by and the behaviour of the major oil companies. It was understandable that various ministers felt uneasy with this situation and in ministerial council demanded more information. The Cabinet, however, never did succeed in cutting loose from this dependency.

It is striking that the government never had a proper picture of the oil supply, whereas former employees of the Foreign and the Military Intelligence Services believe that they themselves had had a reasonably adequate picture. The services were together reading the coded reports and communications of Shell and Mobil, but apparently this information never found its way to the Cabinet.[90]

In this situation, mutual relations between the departments involved were not always optimal. The Ministry of Economic Affairs seemed to have a monopoly command over data relating to oil supplies. Subsequently, dissatisfaction was expressed within the Foreign Ministry over the lack of information that had been passed on by Economic Affairs. This lack of information, as a later Foreign Ministry report rather charitably put it, was partly the consequence of the Economic Affairs apparatus being over-taxed, but at the same time 'Economic Affairs did not keep to the arrangement that talks between their officials and the oil companies could also be attended by Foreign Office officials'.[91] Westerterp also expressed the view later that the information on oil stocks and supplies passed on by Economic Affairs had not always been adequate.[92] Unquestionably, this lack of effective communication between departments played a damaging role.[93]

In the rationing phase, the Ministry of Economic Affairs and Lubbers lost their leading role as initiators in matters and measures of limiting consumption. The car-free Sunday had been the result of a mandate given Lubbers by the Cabinet to take whatever action the circumstances demanded. With the Enabling Act and rationing, the oil crisis had taken a turn which made it the concern of the entire Cabinet in which Boersma and, especially, Den Uyl increasingly played the more prominent roles, while Lubbers himself increasingly came to entertain doubts. These shifts were not conducive to decisive or adroit action, and the effectiveness of the policy suffered badly.

7
From Copenhagen to Washington

In this chapter we shall return to the international aspects of the oil crisis, to the stage after the failure of the European Summit of December 14 and 15 in Copenhagen. At that time, the predominant feeling in The Hague was of great uncertainty over the oil supply. Although it was clear by then that some oil companies were trying to share out the oil between European countries as equitably as possible, it was felt that the oil supply would remain uncertain for the immediate future. Seen in retrospect, December 1973 was the worst month for the oil supply. This was the time that the decision was taken to introduce rationing and to submit the Enabling Act to Parliament. By the end of the year, however, it was becoming clear that the worst was already over and that the supply of oil would recover within the foreseeable future.

The Cabinet, in spite of everything, was taking a more business-like – even pugnacious – approach to its judgements of the oil crisis and the embargo, although its Middle East policy remained low-profile. All mention of the Arab-Israeli war was to be avoided for the time being, all public declarations suppressed, as the Arab countries wished. Any hope of European solidarity had been discarded, and even more: the Cabinet, as we saw, had in the meantime reached the conclusion that *sharing* was, in the end, probably not to The Netherlands' advantage. The national energy situation was in all respects in reasonable shape and more favourable than in several other EC countries. For this reason, the failure of the Copenhagen Summit was certainly not lamented in The Hague.

American Leadership

This chapter deals with all the political twists and turns from the Copenhagen Summit till the Washington Conference called by President Nixon, which opened on February 11, 1974. During these weeks the American government began to get more actively involved in the politics of the oil problem. In the first weeks after the outbreak of hostilities, Kissinger had mainly concentrated on ending the state of war in the Middle East. This was no easy matter. American attempts were made more difficult by the fact that elections in Israel had considerably strengthened the political right, leaving the government of Golda Meir much less room for manoeuvre. Nevertheless, negotiations were opened on December 21, although Syria took no part. Discussions proceeded with difficulty, indeed they were initially adjourned, yet despite this, they led to the signing on January 18 of an Israeli-Egyptian accord guaranteeing disengagement.

Washington's relations with the West European countries at the beginning of the 1970s were not unproblematic. As already discussed in Chapter 3, a number of problems in the area of monetary and trade policy, not to mention the Vietnam war, stood in the way of easy cooperation. It is not so surprising therefore that the first weeks of the oil crisis should witness such a marked lack of harmonious cooperation between the USA and its European partners. Paris, in particular, could not come to terms with the American role in negotiations for a settlement of the October War. Kissinger, on the other hand, had no time for European interference. France tried to get the EC to take a position more independent of the USA in relation to the Middle East conflict; but the EC proved too divided to play any consistent role. In part, this was due to opposition from The Hague. The Dutch Cabinet firmly believed that the EC should not thwart Kissinger's efforts to reach a peace accord, efforts to which much energy had been devoted. At a meeting held in Geneva in December 1973, chaired jointly by the USA and the Soviet Union, the main stumbling block was the question of the participation of an independent Palestinian delegation. The meeting was adjourned shortly after the opening ceremony.[1]

Kissinger's address of December 12 marked the moment when the USA tried to establish its position at the head of the Western world in the problematic area of oil and energy needs. Kissinger called on the OECD countries to develop joint initiatives regarding oil production, the use of energy and the development of alternative energy sources. He proposed setting up an energy-action group to discuss these proposals further. It was this initiative which led to major changes in the whole context in which The Netherlands' oil policy was determined. This was already evident in Copenhagen.

The Den Uyl Cabinet, especially Van der Stoel, had immediately welcomed the American initiative. American opposition to French efforts toward an independent European Middle East policy could only be in the interest of the Dutch, it was thought. Furthermore, commercial rivalry between the consumer countries had intensified since Copenhagen, with various countries attempting to conclude bilateral agreements with the Arab oil-producing countries with the aim of guaranteeing longer term security for their own oil supply. This increasing tension and the politicising of the oil market were remarked on in The Hague in fiercely critical terms. It was hoped there that American leadership might help put an end to this disarray.

Production Limits and Oil Prices

Although the oil price rises had elicited great concern in The Hague from the very beginning, it was initially the production limits that caused the most disquiet. These limits and the embargo seemed to threaten The Netherlands most directly. In October the OAPEC countries had announced a monthly reduction in oil production of 5%. An extra 5% was added to this figure at the beginning of November, as a result of which November oil production in the Arab world would be some 25% less than in September (if we include the embargo against The Netherlands, Denmark, Portugal and the USA). Moreover, it was decided at the beginning of November that production in December would also be reduced by a further 5%, though this further reduction would not be put into effect against 'friendly' states. On November 18 the OAPEC countries announced that, out of appreciation for the EC communiqué of November 6, the EC countries (apart from The Netherlands and Denmark) were exempted from the limit set for December. But at the beginning of 1974, a general restriction would again be imposed, a decision confirmed at a meeting of Arab Oil Ministers over December 8 and 9.

At the end of December it was clear that the character of the crisis had begun to change. From that moment, it centred more and more on prices and increasingly less on production limits. On November 19 the negotiations in Vienna between the OPEC countries and the oil companies had become deadlocked. The OPEC countries had subsequently announced that they would in future set the oil prices unilaterally. On November 22, Wagner had sent a letter to Den Uyl expressing his great alarm at the price rises. On November 30, Dutch diplomatic representatives were fully informed of Wagner's apprehensions via the Foreign Ministry. Van der

Stoel explained that, although the government shared Wagner's view, in view of the embargo it would not seem to be the responsibility of The Netherlands to take any specific initiative within the OECD.[2]

On December 23, the Arab Oil Ministers, meeting for an OPEC conference in Tehran, let it be known that they wanted to double the price of oil. They decided to recommend to their governments to raise the price of the so-called light Arab crude oil to around $11.60 per barrel. Several days later, the other OPEC members followed suit. On December 25 the OPEC countries announced that the production limits then in operation would again be relaxed. All measures previously announced now lapsed, and in addition, the restriction that had reached the 25% figure was now relaxed to 15%. The oil flow would resume, but at a price three times higher than before the war.[3]

The Western countries reacted with shock to these spectacular price increases, even the USA, which had originally supported the idea of a price rise. As early as December 22, the Dutch Director-General for Political Affairs, Van Lynden, received a worried American *aide-mémoire* over the OPEC plans which would lead to a threefold or even fourfold increase in oil prices compared with the pre-crisis period. Price rises of this magnitude would have highly serious consequences for all users, the *aide-mémoire* observed. For this reason, American embassies in the OPEC capitals had been instructed to express their alarm at these rises at the highest level possible. The Hague was likewise asked to take similar steps.[4]

Van der Stoel, however, reacted coolly to this request. Shortly before the OPEC gathering in Tehran, the Dutch Ambassador in Jakarta had received instructions to approach the Indonesian authorities cautiously and to make clear The Hague's concerns over the rise in oil prices. But Van der Stoel had no wish to make any comparable attempt in other OPEC countries. The special position of The Netherlands, i.e. as the target of the embargo, did not after all allow for The Hague to play an active role in the matter of a price dispute. This was why Indonesia had been chosen as the conduit. Van der Stoel also reacted warily to the American request of December 22. The Dutch Ambassador in Tehran was merely informed in full of the American steps, but was not instructed to support them.[5]

It was not only Washington that tried to get the OPEC countries to show moderation and restraint. As it turned out later, there were also British attempts. On January 2, in conversation with the Dutch Director-General for European Cooperation, the British Ambassador at The Hague said that London was highly alarmed by the price rises that had been carried through. In his view, a positive response should be shown to the suggestion of the Shah of Persia for a dialogue between OPEC and the OECD. The British government proposed a formal approach to the OPEC countries to appeal for reason and moderation. Anxiety should be expressed over the December 22 price decisions of OPEC and the Gulf States, and it should be requested that these be introduced in stages. Finally, any new destabilising decisions at the coming OPEC meeting on January 7 should be prevented.[6] It would be helpful if as many OECD countries as possible would support this effort, the British Ambassador thought. After discussing this internally, the Director-General reported back that The Hague was of roughly the same opinion and supported the British plan.[7]

Evidently, bilateral steps were going to be of little help for the time being: talks between producers and consumers were required. But there did seem to be common points of contact. At a press conference on December 23, the Shah had appealed for a dialogue between OPEC and OECD, which should take shape within the next six months. The OPEC Secretary-General, Abderrahman Khene, had also expressed the same kind of idea. The appeal was endorsed by several Western governments, including The Hague, where it had been realised that these price rises would have enormous consequences for the Dutch and for Western economies in general. Particularly within the Dutch Directorate-General for International Cooperation, there was also increasing concern for the position of the developing countries.

The Embargo

However much the crisis was beginning to change in character, the embargo was still in force. Throughout the latter part of December and in January, the Cabinet took a very cautious position on the Middle East conflict: theirs was still a 'low profile' policy. In any case, The Hague declined to make public pro-Arab pronouncements and as a consequence came in for some criticism. The conservative-liberal VVD reproached the government for doing too little to defend the Dutch national interest in the Middle East (as Brussels had succeeded in doing by giving its own interpretation of the EC declaration). Van der Stoel was accused of conducting 'bystander politics' which simply ignored the economic interests, whereas other West European countries did not hesitate to act blatantly in whatever way necessary to secure their oil supplies from the Arab countries.

The Minister also received various suggestions, even from his own diplomats, that he might try to meet the Arab countries halfway. The Dutch Ambassador in Bonn, De Beus, for instance, conveyed the Egyptian suggestion that Den Uyl might once again confirm the November 6 EC declaration in an interview. A separate communiqué would be best, but it was understood in Cairo that this was not feasible: an interview would perhaps be a satisfactory solution. The German Chancellor Brandt also had a similar conversation with a Syrian journalist. This journalist was prepared to arrange in detail questions and answers so that Den Uyl need not be drawn any further than was possible. De Beus hoped that the Premier would permit such an interview since, after all, even if the worst came to the worst, there was nothing to lose.[8]

This plea from De Beus was not unsupported; throughout the entire crisis there had been suggestions from a number of Dutch diplomats that the government should, in some way or other, do more to accommodate the Arab countries. Similar suggestions were heard from the Directorate-General for Political Affairs, including from Director-General Van Lynden himself. But Van der Stoel remained resolutely opposed to such a public declaration on the basis that any such declaration would only invite new demands.[9]

Moreover, The Hague's position became no simpler when the Arab countries decided in Kuwait on December 23 to reward some countries for their friendlier attitude. It is possible that a certain reconsideration of the economic effects of the earlier sanctions played a part in this. In any case, it was decided to consider Belgium as a 'friendly country' and thus to give it a more favourable treatment. It certainly seemed like a reward for the fact that Belgium had publicly taken the line that Israel had to withdraw from all the occupied territories. It was decided, furthermore, to allow oil supplies to Belgium via Rotterdam once again, a decision that was also significant for The Netherlands. On January 4, Van Schelle reported that the Saudis were indeed pleased with the Belgian guarantee that oil intended for Antwerp really would be pumped to Antwerp via the Rotterdam-Antwerp pipeline. The quantity and quality of the crude oil arriving in Rotterdam for this purpose would have to be determined by attested Belgian shipping inspectors.[10] Japan, one of the most oil-dependent countries in the world, was also rewarded. Despite heavy American pressure, the Japanese government had already publicly stated on November 22 that Israel must withdraw from all the occupied territories.[11]

But The Netherlands was not considered to be 'friendly', nor even neutral. This was evident toward the end of December when Great Britain, among other countries that hitherto had hardly or not at all supported The Netherlands, appeared ready to undertake diplomatic activity to try to get the embargo lifted. At the end of 1973, it was evident in The Hague that the British embassies in the Middle East had been instructed to make efforts to get the oil restrictions relaxed and, further, to argue the Dutch case. This effort, however, bore little fruit. The Egyptians perhaps recognised that the time had come for the Arabs to change their attitude toward The Netherlands, but they were not yet ready to undertake any concrete step, in that direction. Elsewhere, it was clear that British efforts in Kuwait had accomplished nothing, while the prevailing judgement in Syria and Libya gave little cause for optimism.[12]

It became clear after several weeks, however, that the problem lay mainly in Saudi Arabia. Yamani may have admitted that the embargo against The Netherlands and the USA was completely ineffective,[13] but King Feisal was not to be persuaded to lift the embargo. His stubbornness, according to the Dutch Ambassador Derksen, could only be softened by some new Dutch gesture, which might consist of a reply to the ultimatum of October 1973, delivered by Feisal personally in his talk with Van Roijen and repeated in his letter to Queen Juliana.[14] Little in the way of a favourable result should be counted on from the pressure exerted by other Western countries. As Derksen on another occasion observed, American mediation, specifically through the US Ambassador in Jeddah, had earlier had an adverse effect on the Dutch cause.[15]

Van der Stoel was not convinced by Derksen's argument, and declined to put out any new statement or to send a mission to Jeddah. Wagner, the chief executive of Shell, shared this viewpoint. He stressed on January 11 that the Dutch position in the Arabian Gulf States had not improved, and that there was therefore little point in once again sending representatives to the relevant capitals.[16]

In mid-January, however, there were briefly rumours in The Hague of a possible end to the embargo, fed by remarks of the Kuwaiti honorary consul, Rabbani. On January 14, the latter stated on Dutch television that the embargo had done its job. At the same time, there was now a much better understanding in The Netherlands of the Arab side of the conflict. Rabbani therefore showed himself ready to urge his Arab colleagues to lift the embargo. NRC *Handelsblad* reacted ironically to Rabbani's remarks, and with good reason: within a few days, his suggestions were repudiated by the Kuwaiti government.[17] At the same time it was put about that Rabbani had fallen into disfavour.[18]

Despite this, from other quarters it became evident in January that there was movement on the embargo front. After a tour of the Gulf capitals, the Dutch Ambassador Schorer reported that Abu Dhabi, Bahrein and Qatar seemed to feel more inclined to lift the embargo than to continue with it. The situation in Kuwait was rather opaque, but Schorer thought that

> steadily diminishing communication between our country and the Arab world, a situation that appears to be deliberately encouraged by some influential Arab countries, could have extremely undesirable consequences.[19]

One hears yet again in this report the echoes of criticism of Van der Stoel's policy of passivity.

The hope of a speedy end to the embargo received a boost by the signing in Geneva, on January 18, 1974, of a disengagement agreement between Egypt and Israel. On January 19 and 20, Kissinger subsequently held discussions with the Jordanian King Hussein and with the Syrian President Assad. Although after these talks Kissinger spoke positively of suggestions made by Assad concerning disengagement, there remained problems over prisoners of war. Syria was not prepared to publish a list of the names of Israeli prisoners of war.[20]

Nevertheless, the situation in the Middle East had markedly improved. On January 22 Kissinger therefore declared that the Arab countries should now rescind the embargo measures, which was construed in the Dutch press as a sign that the embargo would indeed soon be over. But at the Dutch Foreign Ministry the possibility had been taken into account from a very early stage that the embargo against the USA might be withdrawn without lifting the embargo against The Netherlands. On January 12 the Dutch Ambassador in Washington was given the job of urging the State Department to do everything possible to get the embargo against the Netherlands ended. Ambassador Van Lynden received an assurance from Under-Secretary Donaldson that Kissinger had the Dutch interests 'very much at heart'.[21] Over the following days, however, some Arab diplomats made it abundantly clear that, should the embargo against the USA be lifted, it would continue against The Netherlands as long as the Dutch Cabinet made no explicit call for a complete withdrawal of Israel from all the occupied territories.

Furthermore, despite all the secrecy over exact figures, it had by this time become clearer that the embargo had been more or less ineffective. In early January there was renewed anxiety in the Hague when it became apparent that Parisian governing circles were convinced that the oil situation in The Netherlands was far less serious than represented by the Dutch government. In this context, some French officials referred to remarks of such Dutch authorities as the Mayor of Rotterdam, W. Thomassen. Pompidou had made similar comments at a press conference. At a New Year's reception on January 3, the French President jokingly remarked: '*On va connaitre la solidarité, car le gaz hollandais va se solidariser avec le pétrole*', referring of course to the price of natural gas.[22]

In fact, the Dutch Ambassador in Paris, De Ranitz, was not at all sure how to deal with such talk.[23] He was instructed to make it clear that it was difficult to judge whether The Netherlands was in a better position than other EC countries or not. The Dutch government had often called for a more adequate system of information within the EC, and The Hague could not be reproached with reluctance to provide comparative information. In general, the Ambassador should be on the alert for optimistic expectations of the future; after all, the embargo was still in force. The government regretted the fact that, because of existing differences within the EC, the French press should so rapaciously seize on reports of the Dutch oil supply turning out better than anticipated. It was also regretted – although this sounded rather hypocritical – that the Dutch media were not themselves free of a certain *Schadenfreude* over the problems of supply that the French were also wrestling with. Referring to the recent price rises, De Ranitz should stress that oil problems had to be tackled communally. The 'autonomous actions' preferred by Paris had only had adverse, e.g. price-increasing, consequences. Besides, De Ranitz could point out that the companies themselves had in fact assumed the responsibility for sharing out the oil. Had this not been the case, The Netherlands would certainly have found itself in a far worse situation. But at the same time, he should point out that the Dutch Cabinet also found it difficult to accept that, in the absence of any agreement between the governments, this role should have to be entrusted to the multinationals. As Den Uyl proclaimed at the end of January: 'It is unacceptable that private enterprise should determine where the oil goes to...'[24]

Nonetheless, by the beginning of January, the oil situation in The Netherlands had not been bad for some time; and as far as other measures taken against the Dutch were concerned, the situation was now beginning to improve. For instance, it was reported from Kuwait that trade and business in the Gulf States no longer experienced any difficulties, although it had to be said that goodwill was in short supply.[25] Matters were not so clear elsewhere. Some Martinair flights, to Tripoli for example, were resumed in mid-January, but KLM had not yet resumed normal services. In Libya, Syria and Iraq, the ground staff still refused to provide ground services for KLM, while in Iraq and Egypt several other boycott measures remained in force. At the beginning of February, the Syrian restrictions against Dutch shipping and air flights would be lifted; those in Iraq and Libya, however, were to continue for some time. All considered, in the judgement of the Dutch Foreign Ministry, the damage suffered by Dutch trade as a result of The Hague's stance in the Middle East conflict remained limited.[26]

Oil for Arms

The situation surrounding the oil supply in Western Europe was complicated in this period from mid-December to February by the attempts of several countries, certainly after the failure of the Copenhagen Summit, to secure their own supply of Arab oil through bilateral agreements. France and Great Britain especially tried to conclude such bilateral agreements, but Belgium, Italy, Japan and West Germany also had a try. These moves usually went hand in hand with political statements which went some way to meeting the Arab standpoint. Furthermore, in the context of such agreements, the oil purchasers appeared ready to accept remarkably high prices. In order to limit the consequences of these high prices, Paris and London tried to strike deals whereby oil would be exchanged for technology and weapons. To the Dutch way of thinking, bilateral agreements of this kind had to be categorically rejected as merely serving to drive up the oil prices.

There had already been reports in The Hague, before Copenhagen, of the bilateral activities of several non-EC states. Madrid, for example, had successfully sent a mission to the Arab oil countries. Saudi Arabia, Algeria and Iraq specifically were to continue normal deliveries to Spain and in some cases even increase them. Madrid had for some time wanted to reduce its dependence on the major oil companies by dealing directly with the producers. The Spanish position with regard to the oil crisis was much like the French, according to the Dutch Ambassador in Madrid. The Spanish government was against any action by the consumer countries that might be construed by the Arab world as 'ganging up'.[27]

Under pressure from their commercial and industrial community, and from Japanese public opinion, the Japanese government also decided to send a mission, led by Vice Premier Takeo Miki, to various Arab countries. Japan was 100% dependent on oil imports for its energy, specifically oil from Iran, which was evident in an Arab-inclined foreign policy.[28] The Japanese mission visited several Arab countries and tried to secure the oil supply in exchange for economic and technical assistance. In January, a Japanese-Iranian agreement would be concluded which guaranteed oil deliveries in exchange for Japanese support in building a joint refinery and other petrochemical projects. Several days later there followed an agreement with Iraq, and at the end of January also with Saudi Arabia.[29]

On December 21, several days after the failure of the Copenhagen Summit, the Dutch Ambassador in Paris reported that France was also engaged in trying to conclude a bilateral deal with Arab countries. The country mainly concerned was Saudi Arabia. However, it was unclear whether any credence should be attached to these 'arms-for-oil' reports, which were also starting to be reported in the Dutch press in early January.[30] According to a French spokesman, the delivery had been guaranteed of 800 million tons of oil over a twenty-year period, in exchange for technological help and military supplies.[31] American sources were reporting that only 200 million tons of oil were involved over a three-year period. This report subsequently turned out to be the more accurate. The Stockholm International Peace Research Institute later confirmed that a deal was concluded in December 1973 whereby France would receive 200 million barrels of oil in exchange for 38 Mirage jets, rockets, AMX-30 tanks, frigates and minesweepers over three years.[32] But the discussions with Saudi Arabia of early February 1974, over a long-term agreement, ran aground.

At the end of January, the Dutch Ambassador in Paris sent word that there was now also talk of an arrangement between Paris and Kuwait, providing for oil deliveries for twenty years in exchange for technological help (in the construction of a nuclear power station). The agreed oil price, it was said, was probably high. It was later admitted by the French that the price was on the high side. In Paris, however, such an agreement was thought preferable to having no agreement at all, when supplies would to a large extent be dependent on the international oil companies.[33] Initially, it was denied in Paris that there was any agreement included in this deal to supply weapons, but of course this was later found to be so. In February there followed agreements with Iran and Libya, always following the same pattern of long-term oil supplies in exchange for technology, nuclear reactors and arms.[34]

In The Hague's view, this approach went hand in hand with radical concessions to the Arab point of view. There was therefore little sympathy to be expected for the Dutch problems from the French side. De Ranitz reported that it was highly unlikely that Jobert, in his travels through the Middle East, would have had anything favourable to say on behalf of The Netherlands during talks with Arab members of government.

> Such would not be wholly in keeping with the idea, also entertained by President Pompidou, that France must now capitalise on their acknowledged pro-Arab policy, so unpopular with the French people but pursued by their government for many years.

The question of the embargo against The Netherlands, in the French view, had to be resolved within the context of a European-Arab rapprochement.[35]

It was not only the French who were trying to follow the bilateral route to securing their oil supplies. The British were at it, too, although they were doing so with rather less ambitious aims than the French, who were advocating a European-Arab political rapprochement. For the British it was solely a question of oil. On January 4, Gevers, the Dutch Ambassador in London, following a conversation with Under-Secretary Taylor, reported that Great Britain and Saudi Arabia were going to conclude a deal involving the supply of thirty million tons of oil per year in exchange for machines and arms. Taylor emphasized that no agreement had so far been reached, but that it could well materialise. At that moment, moreover, a British mission was on its way to Iran. He admitted that this bilateral approach could lead to playing off one industrialised country against another, but the alternative was buying by auction, with enormous price rises as a consequence. Furthermore, the oil-producing countries would shortly assume a far greater share in oil production, as a result of which the Western countries would be forced to negotiate bilateral agreements. The British government absolutely did not want to put the majors offside, but there was now simply no alternative to bilateral oil diplomacy.[36]

Later supplementary reports of British oil diplomacy followed, which incidentally paid more attention to the anti-Dutch embargo than had the French. The mission to Iran had been promising, in spite of Japanese competition, but talks in Saudi Arabia had come to nothing. In the case of Iran, five to six million tons of oil extra in 1974 had been discussed, in exchange for the supply of such basic industrial materials as paper, steel, cement and textile fibres of a total value of 240 million dollars. The Iranian deliveries were achieved through extra production, not to be deducted from normally available quantities. The British appeared not to pay a higher price than was asked of the oil companies, which the French certainly had. This transaction was labelled by the Americans as 'aggressive bilateralism' and condemned as 'naïve and dangerous'.[37] Shell, of course, also regarded all these bilateral activities with cynicism and distrust. The British director of *Shell Transport and Trading*, Frank MacFadzean, ironically observed:

> Delegations and emissaries, politicians and friends of politicians, most of them with little knowledge of the oil business, descended on the Middle East like a latter day plague of near Biblical proportions.[38]

The end of January brought Yamani and Abdessalam on another visit to Western European countries, an occasion of further European-Arab rapprochement, this time involving Italy and West Germany. The West German government also tried to come to some barter arrangement with Arab countries, which included the prospect of the sale of Leopard tanks. This was tied to concessions to the Arab political standpoint. As was later to be revealed, the West German Foreign Minister Scheel, who on January 1, 1974, succeeded his Danish colleague as the new chairman of the European Community, sent a letter to Yamani and Abdessalam which (according to the Dutch Ambassador De Beus) went further than the declaration of the Nine. Bonn may have recognised in this letter the possibility of border changes, but declared that extending territories through the use of force was no longer possible in our time. So long as it remained impossible to reach any other voluntary agreement, a peace settlement would have to be based on a total end to the occupation of foreign territories and thus a complete withdrawal of foreign troops. One had to base any accord on the borders existing before 1967. No peace accord could be possible without settling the Palestinian question, which was not to be understood as purely a problem of refugees.[39] It was afterwards concluded in The Hague that the Bonn government too had been prepared 'to prostrate themselves just to please the Arab countries'.[40] At the end of January, a cooperation agreement was signed between West Germany and Iran worth 2.2 billion dollars.[41]

The Arab Oil Ministers' visit also set the Italian government in motion. The Prime Minister Aldo Moro told the Senate on January 23 that Israel must pay the price for peace and secure borders: unconditional and complete withdrawal from the occupied territories. This was how Rome came halfway to meet the Arab demand for a further, more explicit, standpoint than that voiced by the declaration of November 6, a demand which Moro had up till then refused. According to the Dutch Ambassador in Rome, this U-turn was the result of pressure from Moro's fellow party member Fanfani, who in contrast to the 'European' Moro belonged to the 'Mediterranean' school. Perhaps the oil-grabbing trips of the French and the British had also given Moro pause for new thoughts on the matter.[42] In fact, in February agreements were struck with Libya and with Saudi Arabia. A trip undertaken by Moro to several other Arab countries shortly afterwards, however, was without further success. But in June yet another accord was signed with Iran over cooperation in the field of energy.[43]

The Netherlands emphatically rejected this bilateral oil diplomacy. Such activities not only undermined any possibility of communal policy-making within the EC, as Van der Stoel told the Dutch Council of Ministers, they also thwarted American attempts to reach a peace settlement. European countries, France above all, were now even providing the Arabs with weapons.[44] This did not obscure the fact that interested Dutch concerns were urging a comparable, more active involvement in the Middle East. A highly placed official of KLM thus suggested that Van der Stoel should visit several Arab countries, such as Syria, Iraq, Kuwait and Saudi Arabia, but the latter was as resolutely unreceptive to the idea as usual. There was furthermore another complication, according to Van der Stoel:

> Of course, for my part, were a visit to Arab countries to be considered, I should then also be obliged out of the need for balance to think of paying a visit to Israel, which only complicates the whole affair.[45]

Van der Stoel did, however, consider contacting Arab colleagues in the lobbies of the UN at the coming session of the General Assembly.

The Dutch Foreign Minister thus maintained his reserved stance, certainly reserved in comparison with most of his EC counterparts. It was not only some of those whose economic interests were at stake who queried this attitude; in the Council of Ministers, too, doubts could sometimes be heard voiced over Van der Stoel's categorical rejection of bilateral diplomacy. For example, on February 1, Minister for Economic Affairs Lubbers warned against a too absolute condemnation of bilateral agreements, for The Netherlands too could find itself forced to take the bilateral route in order to secure its oil supply.[46]

In fact, the Dutch Ministry of Economic Affairs was inclined to follow the example of the other EC member states. At this stage, in line with Lubbers's construction of the situation, the Ministry of Economic Affairs was considering the possibilities of improving bilateral economic contacts with the oil-producing countries. But this conflicted with Foreign Ministry policy. Nevertheless, it was on record that various EC countries had already made arrangements to the effect that oil deliveries had been agreed in exchange for 'compensatory transactions', which in the case of Britain, France and West Germany included supplying weapons. In the view of Economic Affairs, a prompt stand by The Hague was essential 'if we were not to wake up and find we had missed the boat, because already some Western countries had stolen a lead on us'. One of the possibilities discussed in this context was the use of development aid, though it was assumed that such an approach would be blocked by objections from Minister for Development Cooperation Pronk,[47] who was indeed in total disagreement with Lubbers in the Council of Ministers. The oil problem, in Pronk's view, had to be resolved in a multilateral framework so that the developing countries could also be involved.[48]

An Invitation from Nixon

In Copenhagen the EC countries could not agree on a common Middle East policy. Neither had it been found possible to put in train a communal policy regarding oil and energy provision. Moreover, several EC countries had by then become entangled in attempts to secure their own oil supply by bilateral negotiations. There were accordingly powerful centrifugal tendencies apparent within the EC; and not only within the EC, but throughout the entire Western world.

Shortly before the European Summit, the American government had called for talks between consumer countries. In his December 12 address to the *Pilgrims Society* in London, Kissinger had warned against achieving European unity at the cost of the Atlantic alliance. There had to be a unified effort by the Western countries to find a common solution to the energy problem. The USA could only with great difficulty realise such a solution through her own power, and an isolated Europe would find it completely impossible. For this reason he had proposed that the USA, Canada, Western Europe and Japan should set up a *Joint Energy Action Group*.[49]

Against the background outlined above, this initiative appeared to be aimed as much against unilateral West European moves as against the development of an omnipotent OPEC cartel (with the concomitant danger of steeply rising oil prices). In a conversation in The Hague between the

Department for Economic Cooperation (DES) and Shell representatives, it was assumed that Kissinger had indeed two concerns in mind, viz. the danger of further bilateral agreements with producer countries, and the risk that the OPEC success would lead to comparative initiatives for other basic materials.[50] Kissinger himself later wrote that the initiative was mainly an attempt to thwart the bilateral activities of certain West European countries.[51]

As he explained in the Council of Ministers, Van der Stoel took a positive attitude to the initiative, despite all the complications. Immediately after Kissinger's speech, he had ordered his civil servants to look into the American proposal. In consultation with the Ministry of Economic Affairs, a mixed workgroup was assembled to work out a course of action. The topic of discussion at the first meeting of this workgroup was more effective international cooperation in the field of energy. There were three areas where this could happen. In the first place, through the co-ordination of national measures to cut back on energy use; secondly, by stimulating the exploration of oil and gas and the development of alternative energy sources; and thirdly, by encouraging producers to expand the supply of oil and gas in order to tide over the interim period before alternative energy sources were adequately available. One option was to get the producers to invest their profits in alternative energy sources, which could also secure their own future once the age of oil was past.

In any case, the workgroup concluded, what had to be prevented was that the American plan should simply sink in the boggy swamps of the EC.[52] In fact, the Kissinger plan rapidly led to open dissension within the EC. At the beginning of January, it is safe to say, the Dutch and British standpoints over Kissinger's proposed 'energy group' did not significantly differ, but it was entirely otherwise with the French. Paris strongly preferred talks between consumer and producer countries in which the poorer developing countries would be directly involved. Kissinger's proposal, on the contrary, came down to discussions between consumer countries, with the producer countries excluded. But the French point of view was not entirely clear: the exact wording of Kissinger's initiative, according to De Ranitz, could be important.[53]

On January 9, 1974, Den Uyl received an invitation from Nixon which included a copy of Kissinger's text with its exact wording. The aim of the conference, Nixon explained, was to construct a common position regarding the energy question. The Netherlands was therefore invited 'as a first step' to a gathering of Foreign Ministers from the 'major industrial consumer nations' that would be held on February 11 in Washington. The

Ministers' job would be to put together a *task force* 'drawn from the consuming countries which would formulate a consumer action programme'. The objective was the solution of those problems that had arisen as a result of the explosive growth of the world-wide energy demand, including the search for alternative sources; in addition a 'concerted consumer position' had to be developed

> for a new era of petroleum consumer-producer relations which would meet the legitimate interests of oil producing countries while assuring the consumer countries adequate supplies at fair and reasonable prices.

The interests of the developing countries must at the same time also be kept in mind. After ninety days, a further meeting of 'consumer and producer representatives' should be held. Nixon mentioned that a simultaneous letter had been sent to the OPEC capitals, announcing and elucidating the American proposal. In the first instance, among the EC countries only Great Britain, West Germany, France, Italy, Norway and The Netherlands had received an invitation, as well as Canada and Japan. Bonn, charged with the chairmanship of the EC, was asked how the other countries should be represented.[54]

In The Hague, a session of the Foreign Ministry-Economic Affairs workgroup was held on January 9 to discuss Kissinger's call for multilateral talks, a meeting which concluded that Nixon's invitation should receive a positive response as promptly as possible. The workgroup had meanwhile come to the conclusion that the objectives of international talks in Washington had to be as pragmatic and as concrete as possible. The workgroup argued

> that the consumer countries represented there should not attempt to realise a complete, fully worked-out standpoint regarding the energy crisis, since this would take much too long.

Consultations with the producer countries should be begun promptly, because this could possibly 'put a brake on the trend of consumer and producer countries negotiating separate deals, as well as autonomous actions by producers.' In addition, developing countries should also be drawn into these talks between consumer and producer countries as soon as possible.[55]

On this point, there were considerable differences of opinion within the EC. Some EC countries, France specifically, were first of all striving for negotiations between the EC and the producers. The Netherlands wanted such talks to follow from the Washington Conference, so that the USA would also be present. As far as the EC's role was concerned, it had to be a case of translating global arrangements into communal EC policy. But, as was clearly realised by the Dutch Foreign Ministry, this principle was entirely counter to French policy.[56]

Despite all differences of opinion, an attempt was made to reach a common EC position at the Washington Conference. In preparation for this conference, the European Commission made a number of proposals, but the Dutch Foreign Ministry doubted whether such EC discussion was necessary or sensible. There was growing resentment within the Ministry over the role of some EC partners. After all, the EC partners had seen no need for joint consultations in their bilateral dealings over oil.[57] Van der Stoel shared this scepticism. To put this surliness of the Dutch attitude into context, it has to be remembered that at that time not all EC countries had been invited. Only the *major consumer countries* had been approached by Nixon. Only at a later stage would all the EC member states be involved.

On January 10 the Ministerial Council for European Affairs discussed the question of possible EC preliminary consultation. Van der Stoel warned of American frustration if the EC countries responded to Nixon's invitation by stalling until they could first reach a common standpoint. The Netherlands should straightforwardly accept the invitation. However, Under-Secretary for European Affairs Brinkhorst thought that The Netherlands would place itself in an incredible position if it by-passed communal EC deliberations, a point of view endorsed by Den Uyl. The Council for European Affairs voted accordingly.[58] On January 10, various diplomatic outposts were informed that the Cabinet welcomed Nixon's initiative. The Hague shared the President's view that the Washington Conference should be followed promptly by talks between consumer and producer countries, talks in which the developing countries should also be involved.[59]

This was the standpoint more or less agreed on. However, during the Dutch Council of Ministers of January 11, it emerged that Minister Pronk had principle objections to the Washington Conference, which he considered a manoeuvre to exclude not only the oil-producing but more especially the developing countries. He preferred direct talks between consumers and producers, in which the developing countries must be involved, as advocated by the French. Despite Pronk's objections, however, the Cabinet agreed to accept Nixon's invitation.[60] Given the wording of

Van der Stoel's earlier circular, there was in fact no option.

A meeting of EC Foreign Ministers was held in Brussels on January 15 at which the Nine were to work out a common standpoint for the Washington Conference. At first, it looked as though agreement would be unlikely: Belgium and Denmark both had reservations over the American initiative because, like Ireland and Luxembourg, they had not been invited. As regards this criticism, all EC Member states plus the OECD Secretary-General were subsequently invited. During the discussion, Jobert initially resisted a joint acceptance of the American invitation, which hardly came as a surprise, since the French government had already indicated that it found the meeting in Washington neither expedient nor desirable. It was clear that in several respects the basic assumptions of the conference were in conflict with the French Middle East and oil policy.

In Brussels, Van der Stoel's stance was one of strong support for the American initiative. Rather opportunistically, he observed that the invitation could have a very positive effect in stimulating the search for a common EC energy policy. Of course, it was not so simple; but after laboured discussions agreement was eventually reached. All EC member states should be able to participate in the Washington Conference, and in addition the chairmen of the EC and the European Commission, Scheel and Ortoli, should also participate, on behalf of the EC as a whole. The intention was that the Committee of Permanent Representatives should work out a common standpoint on the basis of proposals from the Commission. That was to prove difficult. Jobert remained fiercely opposed to the Washington Conference and said as much in a letter to Kissinger. The French attitude to the plan was subsequently described by Kissinger as a 'lack of enthusiasm bordering on hostility'.[61]

French Obstruction

As a result of French opposition, but also because of the sometimes less than thorough preparation on the part of the Americans, it remained for a long time unclear what exactly was going to happen in Washington. On January 18, a conversation took place in The Hague between Foreign Ministry officials and representatives from Shell, who reported that there were by now some seven working groups set up to look into the various aspects of the oil crisis. Their reports were to be ready by January 23 and would provide the groundwork for a round trip by Under-Secretary Donaldson, who would inform the participant countries as to the agenda.[62] Even the Dutch embassy in Washington sent word back to The Hague of the intended Donaldson trip; but in the event, it never took place.

On February 1, Van der Stoel received an aide-mémoire from the US Ambassador in The Netherlands with the proposals for the Washington Conference, which were in fact already known in The Hague in outline. On February 8 a preliminary meeting of *high officials* would be held under Donaldson's leadership. With reference to the American proposal to discuss the economic and monetary aspects in separate workgroups, Van der Stoel said that Finance Minister Duisenberg would willingly come. Ambassador Gould emphasized that his government would find it regrettable if at this stage, through informal contacts with Arab countries, basic principles began to be established that anticipated the outcome of Washington. His government, he said, attached great importance to avoiding ruinous competition. These were known views, thoroughly endorsed by Van der Stoel. He pledged to prevent damaging EC initiatives, but warned that other EC member states had other ideas on this.[63]

It also emerged from the American aide-mémoire that no immediate discussions between consumers and producers, so devoutly desired by The Hague, were on the agenda. Van der Stoel spoke his mind on this in Dutch ministerial council, complaining more generally of the inadequate preparations. He had been unable to get any clarity on the agenda of the energy conference. On the one hand, it was about analysing the energy situation, on the other, a question of what could be done between countries with energy problems in order to master those problems. There was also uncertainty over the role of the EC. Paris had not yet responded, and it was very much in question whether the EC could come forward with a common standpoint.[64]

Indeed, it turned out to be very difficult to formulate a common standpoint. This was already evident during the discussions of the Committee of Permanent Representatives. The Dutch PR had received instructions to strike a positive attitude and especially to emphasise the need for reasonable and stable price developments, and for this reason the necessity of refraining from bilateral actions that drove up prices. The French PR, however, took a contrary stance, insisting that the main issue for discussion were the objectives of the Washington Conference. He declared that in his view the problems were of a world-wide nature and therefore should be discussed in a world-wide context. In the meantime, France had therefore taken the initiative of raising the problem for discussion within the UN.[65]

The French PR also stressed that the Community had its own responsibility with regard to relations with producer countries. This appeal for an independent European energy policy was going too far for London, which up to that point had supported Paris. The British PR, Michael Palliser, emphasized that the great problem of energy could not be resolved without cooperation with the USA. London was therefore no advocate of a European 'go-it-alone' policy.[66] It was clear that Great Britain and France had reached fundamentally opposite positions on this issue.

Subsequently, the Dutch Ambassador in Paris, De Ranitz, reported that the French Director-General for Political Affairs, Puaux, believed that Western Europe needed to make itself independent for its own energy requirements, which in the first place would have to take place through closer European-Arab cooperation. The French obstruction of the Washington Conference was partly aimed at making clear to the Arab countries that Europe distanced itself from American policy in the Middle East. Puaux considered the whole meeting part of a new American plan to re-exert its dominance over Western Europe again, after the failure of its 'Year of Europe'. As far as talks with the producer countries were concerned, Puaux confirmed that France had in the meantime contacted the UN Secretary-General, Waldheim, with the proposal to raise the oil problem for discussion under UN auspices.[67]

These developments within the EC were met with cynicism in the Dutch press. *Trouw* commented that rough play was increasingly getting the upper hand.

> The European Community has already been playing in injury time for the past month or more: an unmistakable sign of European rough play ... – It all raises the strong suspicion that the term 'European Community' is gradually becoming little more than a flag on a farm barge sinking in a pool of nationalism and self-interest.[68]

On February 5, nonetheless, a communal mandate was drafted for the EC delegation, in which ample concession was made to the French objections. Thus, no permanent consultative arrangement should be set up between consumer countries, as the Americans wanted. A confrontation between consumer and producer countries must at all cost be avoided. The tone of this French-oriented mandate clearly indicated a preference for resolving oil problems through direct contact with the producer countries, for which purpose a permanent dialogue needed to be established.[69]

Van der Stoel remarked afterwards that the result actually suited him better than he might have expected, given the negative French attitude.

Although the effort to reach bilateral agreements was not clearly repudiated, there had been agreement that countries should not take separate measures that could seriously damage international trade. And although the conference must not become a permanent consultative body, the possibility of working groups remained open.[70]

All in all, the Cabinet was generally positive over the American initiative. There were, of course, second thoughts over the Nixon government's attitude, as on a number of other grounds (finances, trade, etc.). Such thoughts were voiced by the Dutch OECD Secretary-General Van Lennep, who confessed himself pessimistic over the conference and had doubts about the way that Nixon was fulfilling his executive role. He thought the invitations had been sent out without any prior consultation; the climate had only been worsened by Kissinger's criticisms of the French, although in his judgement the American presentation required a solid consumer front; and furthermore, the US approach had so far bypassed what Van Lennep saw as essential: the immediate beginning of discussions with the oil-producing countries over the long-term substitution price for oil. Finally, he had the impression that there was a bilateral American action afoot, with the help of private banking, to stimulate investment in the USA. This could be disastrous, since it only made the *recycling* of oil capital to countries most in need of it more difficult.[71]

Deliberations in European Political Cooperation

The American attempt to take the lead once more in the field of energy provision had its impact on the Dutch government's attitude toward the oil crisis. This was evident, for example, in their handling of their own proposal for providing aid to the Middle East. This plan, of which there is scarcely a trace to be found in the Dutch Foreign Ministry archives, was undoubtedly intended, at least in part, to win Arab appreciation. It had been briefly raised during the Summit, but it had been decided in short order to refer it to the *Comité Politique* (of European Political Cooperation). After Copenhagen, doubts began to gain currency at the Foreign Ministry as to whether it was so sensible after all to press ahead with the proposal. Van der Stoel stated that, in any case, the American initiatives had to be taken into account when presenting the proposal. It should not be seized on simply to frustrate European cooperation. At the Foreign Ministry it was decided on January 4 to defer publication of the plan until February. The Dutch embassies in EC member states were informed that the Director-General for Political Affairs, Van Lynden, would suggest in the *Comité Politique* that the proposal be postponed.[72]

But it was not such a simple matter to drop the plan. Furthermore, it emerged that there were considerable differences of opinion within the *Comité Politique*, specifically between The Netherlands and France, over the details of the plan. The French DGPA proposed limiting cooperation to the Arab countries and excluding Israel. In addition, consultations should have a predominantly political character. The *Comité Politique* drafted a compromise that would go some way to satisfying both Dutch and French conceptions. Even so, there were serious qualms at the Dutch Foreign Ministry, because the plan now threatened to pre-empt the outcome of the Washington Conference. It was decided to ask the European Commission to develop the aid plan further, which would gain a bit of time. Furthermore, 'non-Arab countries' also had to be involved in the cooperation between the Nine and the Middle East.

The Hague made no more fuss about the plan, though of course it remained an attractive proposition to win favour in the Arab world. On January 21 Van der Stoel gave the Dutch diplomatic posts in the Arab countries permission to make known the outlines of the plan where occasion arose. Ambassador Bentinck immediately warned that this could backfire, as the plan referred to the whole region and could therefore be considered as an attempt to provide support for Israel via a roundabout route. It was a warning the Dutch Foreign Ministry took to heart.[73]

The question of the aid plan became increasingly dominated by the French effort to bring about a dialogue that would be both economically and politically oriented, an effort regarded by The Hague with great distrust. But this notwithstanding, the French managed to achieve an important success in January in the context of European Political Cooperation. During a session of the EPC Middle East working group, it emerged that West Germany was prepared to support the French proposal for joining a European-Arab dialogue. Paris wanted to begin at the level of experts and later to let this turn into a conference of Foreign Ministers.[74] Bonn probably hoped by adopting a more accommodating attitude to get Paris to participate in Washington. The Netherlands, however, like Denmark, had serious objections to the French proposal. According to The Hague, such a dialogue could damage already existing forms of cooperation, as well as the peace talks in Geneva. The Netherlands argued for this affair to be dealt with in the ministerial discussions of February 14 and 15, which in effect meant postponing it until after Washington. And furthermore, it meant taking the matter out of the official framework of EPC.

But most EC members followed a proposal from the European Commission, which argued for a *joint venture* of the EC and EPC, in which the EC would take care of the economic content and EPC would look after the political share. The Netherlands was the sole member state with objections to such a construction, while most others were very much in favour of getting on with it and saw no problem in simultaneous preparations for both the Washington Conference and the European-Arab dialogue. On February 7, the Dutch DGPA Van Lynden reconciled himself to the majority view, but not before a certain qualifying spin had been introduced into the decision, which implied that it had avoided 'the Nine already committing themselves too firmly at the present stage to modalities and timetables'. To the displeasure of The Netherlands, the suggestion that the American government be informed of the planned dialogue was blocked by Paris.

Once again, The Netherlands found itself playing a tricky Atlantic role. As we saw earlier, the American Ambassador Gould had made it clear to Van der Stoel that his government was concerned over the development of European-Arab relations. At the time, Van der Stoel had said that he would resist any attempts to thwart the conference in Washington. This would prove to be no idle promise, although The Hague was in no position to block the European-Arab dialogue, which would have done the Dutch reputation no good at all – it had been discovered earlier that the contents of EPC discussions were promptly leaked to the press. But at least The Hague contributed to the postponing of the decision. Much to the Dutch satisfaction, the EPC meeting of February 14 was adjourned until March 4.[75]

The Washington Conference

There was a great deal at stake, but the outcome of the meeting would be disappointing. The conference took place under poor auspices. In general, American-European relations at the time were not exactly characterised by high mutual esteem; while Jobert had on various occasions made it quite clear that he had little confidence in any useful outcome.[76] The reaction in most Middle Eastern countries was, predictably, to repudiate the conference. The gathering was seen, especially by the more radical countries like Algeria and Libya, as an American attempt to organise a block of rich consumer countries against the oil-producing countries and the Third World.[77] Algeria had in the meantime proposed dealing with energy problems in a special session of the UN General Assembly.

A heavyweight Dutch delegation made the trip to Washington, consisting of Van der Stoel, Duisenberg and Lubbers, as well as several high officials from the Ministries of Foreign Affairs, Economic Affairs and Finance. It was uncertain till the last moment whether Lubbers would go, largely because of his involvement in a road accident in The Netherlands.[78]

The conference was opened by the hosts. In his opening address, Kissinger proposed setting up a 'semi-permanent' co-ordinating group to see to preparations for a second consumers' conference. At a subsequent stage, a world energy conference that also involved the producer countries would have to be convened.[79] Van der Stoel was the first non-American invited to speak, undoubtedly an intentional gesture of appreciation for his stance.[80] In his address, during which he only incidentally referred to the embargo, he pointed out the scale of the consequences the oil crisis had had. Solutions had to be found that were acceptable to all countries. Although an increase in oil prices was justified, an escalation of prices could destroy the entire structure of the international economy. And that was not the aim. The developed countries must resist the temptation to think too much of themselves. The Netherlands was therefore an emphatic opponent of bilateral oil diplomacy. Some way of finding a common solution must be found 'to increase the flow of assistance' to the non-oil-producing, developing countries. Van der Stoel's address signalled support for the American plans, but not in their entirety. In fact, he specifically welcomed the Algerian plan for a special session of the General Assembly, and he omitted any supportive reference to Kissinger's proposal for a follow-up consumers' conference.[81]

After this, the conference progressed with difficulty, first and foremost caused by the surly, uncooperative attitude of the French. The signs were initially not so grim; the French government had at the last moment decided to let Jobert set out, although Pompidou had forbidden his Minister of Finance, Valéry Giscard d'Estaing, to accompany him. Jobert, whom Pompidou referred to as 'mon Kissinger à moi', tried to prevent any form of practical decision-making and refused to have anything to do with a follow-up conference.[82]

The European split on the issue of energy now came into the spotlight. The British standpoint had in the meantime moved in the American direction. West Germany, then holding the EC chairmanship, was also prepared to go a long way to meet the Americans. Increasingly, France was finding itself isolated.[83] Jobert accused his European colleagues of adopting a far too docile attitude toward the Americans and at one point even saluted them with '*Bonjour les traitres*'. No punches were pulled within the closed doors of the internal EC consultations, the clashes between Jobert and Helmut Schmidt being particularly bruising.[84] The West German Minister of Finance, who because of Scheel's role as EC Chairman was leading the West German delegation, took an outspokenly Atlantic position, emphasizing that relations with Washington were more important for Bonn than those with the European Community.[85]

Jobert refused at the end to sign the final communiqué, but nonetheless voted for the setting up of a *high-level group* to implement the various agreements reached. These included further study of the global energy situation, cooperation in monetary and economic areas, and 'the development of a cooperative multilateral relationship' between producer and consumer countries. This high-level group, soon to be known as the *Energy Co-ordinating Group* (or ECG) had furthermore the task of making arrangements for a new conference, involving both consumer and producer countries.[86] Van der Stoel considered the result an American success; but the Americans had had to make concessions, as the communiqué showed. There would thus be no follow-up conference to be held solely between consumer countries, as Kissinger had wanted; and the Algerian initiative for a special session of the UN was welcomed, partly due to the insistence of the Dutch. This was something else Kissinger would rather not have seen.[87] Where the Americans did chalk up a result, however, was in sharpening the oppositions within the EC. Nixon and Kissinger had aimed 'to shock Western Europe back into line behind Washington', although in this they had only partly succeeded.[88] Certainly, France's position had been weakened.

In the Dutch Council of Ministers, Van der Stoel later described the conference as mainly of 'procedural' character, noting that Jobert had played an obstructive role. The French, of course, had wanted to keep open the possibility of bilateral agreements. Paris had also tried to ensure an exclusive role for Europe in the Middle East, while the USA, as the Dutch Foreign Minister put it, wanted 'global consultation'. In any case, The Netherlands had avoided giving any impression of heading for a confrontation with the Arab countries. Van der Stoel had for this reason also been positive over the Algerian proposal to raise the whole oil problem for discussion at a special session of the UN.

This latter remark about the Algerian proposal aroused differences of opinion in the Council of Ministers. Pronk, with support from Den Uyl, stressed the importance of the special session of the General Assembly.

Lubbers disagreed. In his view, the oil-consuming countries would do better to organise themselves. He was therefore not so keen on a special UN session where, he felt, because of the nature of the context, the oil problem would be pushed into the background.[89]

After the conference, 'Washington' was extensively analysed in a Dutch Foreign Ministry note sent both to the Council of Ministers and to several diplomatic posts abroad. According to this evaluation, the American initiative was intended to strengthen the position of the consumer countries; yet at the same time the conference had had a strongly political character. Four distinct political aspects were identified, There was firstly 'the European-American relationship', the subject of an intense struggle within the EC, specifically between France and the other member states. Paris saw the summit as an American attempt to gain more influence over the energy policies of the EC countries and over their relations with the Arab oil producers. Jobert fiercely resisted these attempts, which if successful would be at the cost of European independence.

> In this controversy, faced with a choice between a European unity which in this field scarcely exists and accepting the cooperation offered by the USA, France's partners in the event opted for the side of the United States. The West Germans above all, through Scheel and Schmidt, made these political aspects a central issue.

This was not to say that the remaining eight member states agreed with all the American proposals in Washington. At the Dutch insistence, the initiative for a special session of the General Assembly was welcomed in the final communiqué, although 'Kissinger was only reluctantly prepared to accept this'. But taken as a whole, on this issue the outcome of the conference was most satisfactory for the Americans.

The second aspect concerned 'relations within the EC'. The EC countries had gone to Washington with a mandate that was both unclear and incomplete. As a result, it had been once again evident that, as long as there was no advance toward internal integration, the Nine were in no position to conduct any clear foreign policy in the wider world. The conference had only served to intensify the crisis atmosphere in Brussels. The third question was 'the relations with the Arab oil producers'. No-one desired a further confrontation with them, and for this reason a further ministerial conference involving exclusively consumer countries, such as the Americans had wanted, was undesirable. To what extent the Arabs would consider the proposed follow-up conference of producers and consumers a confrontation was unclear. In any case, it was highly uncertain that they would be prepared to participate in such a follow-up.

Finally, there was 'the position of the developing countries'. The Washington participants had all pledged to exert themselves seriously to maintain and to expand development aid. The question, however, was how the developing countries could be involved in an effective capacity in the follow-up talks. It was in this context that the progress of the proposed General Assembly was so important.

The conclusion was that The Netherlands positively valued the tasks assigned to the Energy Co-ordinating Group, which could be expected to provide an effective framework for consultation between like-minded consumer countries. It was regrettable that the EC as such was unable to contribute to this consultation process so long as Paris remained aloof. The Netherlands had to continue its efforts within the Energy Co-ordinating Group to bring about talks between the producers and consumers of oil, talks in which the developing countries must also be involved. These consultations would eventually have to lead to a skeleton agreement that could promote a stable supply of oil to the consumer countries and, at the same time, guarantee reasonable, stable and predictable prices.[90]

Conclusion

In the two months from mid-December to mid-February there were various shifts in the political force field generated by the oil crisis. In addition, there were new, fundamental questions that demanded attention: for example, in what international framework should matters of oil and energy needs be discussed. France tried to ensure that the oil needs of Western Europe should be dealt with under the framework of the EC and EPC, as essentially part of an energy policy that would be independent of the USA, and which would in part be based on a European-Arab rapprochement. When this failed, Paris, just like other EC member states, switched to a bilateral approach to the Arab producers. Among the EC member states, France was also the most radical champion of talks within the UN. Washington, however, had little enthusiasm for the UN option: on the contrary, the consumer countries must first develop their own plan of action before they sat round the table with the producer, and later the developing countries.

The Cabinet, although it had adhered to an Atlantic line during the crisis, now stood divided on this question. It was not so surprising that

Pronk should follow the argument of the more radical oil-producing countries, such as Algeria and Libya, for holding global talks; but the Cabinet as a whole adopted a halfway stance with regard to this question. Kissinger's initiative for consumer talks was supported, but on the condition that such talks be followed promptly by discussions with the oil-producing countries and with the Third World countries, if possible under the auspices of the United Nations.

In the months following the Copenhagen Summit, the question of relations between Europe and America burned with some ferocity. The Netherlands, faithful to tradition, put Atlantic unity first. As became clear, this choice was not based on the usual need for unity in the face of a threat from the Soviet Union. As home country of one of the *Seven Sisters* and having an interest in maintaining the existing relations in the international oil sector, and as the object of attempts to break down those relations, The Netherlands was driven toward the American side mainly on political-economic grounds.

At the same time, it has to be said that the American leadership was not without its opportunism and self-interest, as was equally evident in other policy areas. But then again, the forces of opposition were considerable. Paris declined to acquiesce to American leadership, while various oil-producing countries expressed their dissatisfaction over the Washington Conference convened by Kissinger.

The attitude of the Western countries toward the oil crisis was also an example of a disintegration within the Atlantic world that could also be seen in other areas in the early 1970s. At the same time it was clear that the alternative of an independent European oil and Middle Eastern policy was not capable of getting off the ground because of mutual divisions and distrust. The leading role of America was therefore accepted by the European countries *faute de mieux*, in the face of open resistance from the French, and by The Netherlands with more enthusiasm than by most.[91]

8

Sweating it out

As we have emphasized several times, the oil crisis can be approached from different perspectives. In other words, the crisis had various aspects or, if you will, different levels. One important aspect was the relation between the West and the developing countries of the Third World, which had been affected by the crisis in various ways. On the one hand, rising oil prices threatened some developing countries – particularly the more industrialised among them – with ruinous debts; while, on the other hand, the performance of the OPEC countries on the international stage fostered a new self-awareness in the non-Western world. This was expressed in the appeal by the Group of 77 non-aligned countries for a *New International Economic Order* (NIEO).

This call for reconstruction of the international economy was in part a reflection of the power that various anti-Western movements at the turn of the 1970s had developed, particularly in South-East Asia and South Africa. Salvador Allende's accession to power in Chile also seemed like writing on the wall. Calls for a NIEO were frequently heard in the UN, an attractive forum for the supporters of this cause because there the Western countries had no majority. It was entirely understandable that the non-Western aligned developing countries, particularly the more radical oil states like Algeria, should want to raise the issue of the oil crisis within the UN. What they were trying to do was to tie the oil problem into a broader range of political-economic issues.

Against this anti-Western offensive, the Western countries were divided in their attitude. One might even say that this division further encouraged the growing assertiveness of the Group of 77 as well as OPEC. Division also characterised the response in the Dutch Cabinet to the North-South aspect of the oil crisis. As we shall see, some PvdA members of this Cabinet, Pronk in particular, had considerable sympathy for the struggle to achieve international redistribution of wealth – and understanding, therefore, for OPEC's action in raising oil prices. This approval implied support for the idea of placing the oil crisis on the agenda at the UN, whether or not in some broader context. Others, Lubbers above all, took a more business-like approach and argued for the development of Western cooperation, as agreed at the Washington Conference. The problem, however, as we shall see, was that the attitude of the USA did not altogether evince a loyalty of purpose appropriate to its hegemonic position.

The Dutch position, partly because of the dubious American role, would become no easier during this last phase of the embargo. After long discussions, Abu Dhabi, Algeria, Bahrain, Egypt, Kuwait, Qatar and Saudi Arabia decided on March 18 to lift the embargo against the USA. It was a provisional decision: the question would be reviewed on June 1. The embargo against The Netherlands, as well as Denmark, Portugal, South Africa and Rhodesia, was not lifted.[1] Although the oil supply into Rotterdam was slowly returning to its normal level, this turn of events nevertheless proved a blow, for however ineffective it may have proved to be, the embargo was still capable of damaging the interests of the port. The matter therefore remained of getting the Arabs to lift their punitive action.

This turned out to be no easy task, because the circumstances under which the Dutch policy was necessarily formulated were constantly changing. During this last phase from February to July, the embargo and the coalition behind it seemed to change. Those Arab countries that had in the first place advocated the embargo against The Netherlands were now opposed to it, and vice versa. Algeria, the first oil producer to announce an embargo, now began to argue for its removal.[2] Saudi Arabia, initially one of the more moderate states, now turned out to be the driving force keeping it in place. It is a particularly interesting possibility that the Saudis took this stance in collaboration with the USA.

A Second Letter to King Feisal

From the middle of January 1974, rumours began circulating that the embargo against The Netherlands, unlike that against the USA, would not be lifted. On January 12, the Dutch Ambassador in Washington, Van Lynden, was instructed to request the State Department to try to get the embargo against The Netherlands ended. Van Lynden, as we saw, was at the time assured that Kissinger had the Dutch interests 'very much at heart'. But it turned out during the course of January that for the time being most of the Arab countries had no intention of lifting the embargo against The Netherlands. On January 26, the Dutch embassies in the Arab countries were impressed by the Foreign Ministry that it was by no means certain that the embargo would be lifted at the same time as the embargo against the USA. They were assigned to keep their eyes and ears open, but not to approach the Arab authorities officially: the policy was to remain *low profile*.

Yet the time did seem ripe for some kind of diplomatic initiative. Van der Stoel decided on January 28 to propose in Cabinet that the letter from King Feisal, written at the time in reply to Queen Juliana's missive, should be answered.[3] The letter would of course have to give a degree of satisfaction, but without doing violence to The Hague's position on the Middle East. The missive would need to be sent before February 14, the date planned for the following OAPEC meeting in Tripoli. Asked whether Feisal would appreciate such a reply, the Saudi Ambassador Nowilaty answered positively, albeit in his own unofficial capacity. Ambassador Van Lynden in Washington was therefore instructed to see whether the State Department, in parallel with the letter to Feisal, would be prepared to exercise its influence on Feisal 'in a subtle way' to get the lifting of the embargo against the USA linked to a removal of that against The Netherlands.[4]

But the Dutch Ambassador in Jeddah, who had argued earlier for just such an initiative, warned that his American colleague, James Akins, had so far shown a rather unsubtle manner. It might perhaps be more sensible first to try on one's own behalf; and if this proved unsuccessful, one could always turn to the Americans later. Van Lynden subsequently reported from Washington that Under-Secretary Sisco had reached similar conclusions. He repeated that in various conversations with Arab countries, and especially Saudi Arabia, it was always stressed that any lifting of the embargo should not include the USA alone but also its allies.[5] Nothing could be done, however, about the fact that the US attitude, particularly in Jeddah, was apparently not always favourable to The Netherlands. These reports were a first indication of the remarkable intrigues that would be witnessed in the coming months.

In the letter plus enclosure to King Feisal, sent around February 10, the Cabinet took refuge behind the earlier adopted EC standpoint of November 6. Strictly speaking, there was no question of a separate statement such as the Arab countries desired; yet the EC declaration was quoted and paraphrased in such a way that the Cabinet in fact was balanced precariously on the edge of what was still acceptable to the Second Chamber.

The Cabinet based its position on resolution 242 and on the EC declaration of November 6 which reiterated, crucially, the need 'for Israel to end the territorial occupation which it has maintained since the conflict of 1967'. Furthermore, this document emphasized the right of any state to live within secure and recognised borders. The Cabinet also held that the legitimate rights of the Palestinians must be given 'form and substance' within the framework of a peace settlement.

> This implies that the Palestinians, by some means or other, should be given a say in the matter of determining their own future.

The Hague, according to the enclosed document, sought further to make a constructive contribution to the Euro-Arab dialogue.[6]

This was going quite a way in the direction of further clarification, which Van der Stoel had always refused to give. At the end of January, moreover, he was considering a visit to several Arab countries, even though in earlier ministerial council he had spoken rather dismissively of the 'travel bug' contracted by some of his EC colleagues. In this context, at the beginning of February he inquired of the accredited ambassadors in the Middle East how they rated the *low-profile* policy he had been conducting with regard to the Middle East.

There was evidently positive appreciation and support for this policy, although there had been suggestions in more recent months that more initiatives might be helpful. Ambassador Derksen thought the way this low-profile policy had served the national interest 'could hardly be overestimated'. It had prevented worse things happening and furthermore had made it possible for some countries, such as Saudi Arabia, to maintain their economic ties with The Netherlands whilst turning a blind eye. His considered view was that the time was not yet ripe to change this low-profile approach. The relatively healthy Dutch position with regard to oil stocks meant there need be no hurry to honour the Arab countries with a visit.[7] Derksen was not the only one who reacted coolly to Van der Stoel's plan to arrange a Middle East visit. Ambassador Schorer in Kuwait also seemed to have had doubts: a tour by Van der Stoel was perhaps not such a good idea, but something had to be done to improve Dutch-Arab relations. A parliamentary delegation, for example, could break the ice.[8]

Van der Stoel decided to postpone his trip to the Arab countries. At the beginning of February, the Foreign Ministry considered putting together a fact-finding mission, with representatives drawn from the business community; but it quickly became apparent that the latter preferred to defend their interests through their own channels. Furthermore, it was realised that economic relations between The Netherlands and the Arab region were actually suffering very few obstacles. What was least needed, in fact, was publicity.[9]

King Feisal replied on February 28. He thanked Queen Juliana for her letter, which he had not personally received, and promised to look out for opportunities for cooperation and for strengthening the ties between the two countries. At the same time, however, various recriminations were listed. 'What hurt me,' he wrote:

> is to have seen friendly Holland, alone among all European countries, openly expressing sympathy with Israel, challenging in so doing the feelings of all Arabs, who have a plain and an evident right to their home and country.

Feisal reiterated that the Dutch Cabinet must openly revise its position with regard to the Middle East conflict.[10] All in all, Feisal's reply was on the brusque side; there was no mention of lifting the Saudi embargo.

The Lifting of the Embargo against the USA

It became clear at the beginning of March that the Saudi standpoint, and that of other Arab countries too, had if anything hardened. In the Council of Ministers of March 15, Van der Stoel admitted that he had no explanation for this.[11] By now, several Arab countries had adopted the position that the embargo against the USA should be lifted. It was particularly President Sadat who was aware that this embargo would only frustrate progress on the diplomatic front. At first, the Syrian President, Assad, resisted the idea of lifting the embargo, but under Saudi pressure the Arab OPEC countries agreed in Vienna on March 18 that the embargo against the USA should be lifted. Syria and Libya did not initially acquiesce in this decision; in fact, Libya only lifted its embargo against the USA on December 31, 1974.[12]

As feared in The Hague, the embargo against The Netherlands (and Denmark) was maintained. Italy and West Germany, on the other hand, were promoted to the status of 'friendly' states, which served to highlight further the isolated status of The Netherlands within the European Community.[13] In a response to the Vienna AOPEC meeting, the Dutch government issued a public statement in which it expressed 'disappointment' with what it considered unjust discrimination. The Dutch standpoint on the Middle East conflict did not, after all, differ from that of the Nine. The Cabinet assumed 'that the issue of Arab discrimination against The Netherlands would therefore be raised in the near future within the appropriate European bodies'. How and when was not yet clear.[14]

The Dutch Cabinet does not seem to have been in any way intimidated by the Arab OAPEC countries. Van der Stoel immediately let it be known that there should be no expectation of change in Dutch policy, but this did not obscure the fact that there was a real problem.[15] Following the OAPEC decision, Lubbers expressed his concern over the position of the port of Rotterdam.[16] He had already said in the Council of Ministers on March 8 that although the oil supply had recovered, the maintenance of the embargo was nonetheless damaging to Rotterdam's position in the international oil trade.[17]

A first task at the Foreign Ministry was to find out what reasons underlay the OAPEC decision. Ambassador D.R. Bot reported from Vienna that the reason for maintaining the embargo against Denmark and The Netherlands was that these two countries had up to that point declined to issue a more pro-Arab statement. Italy and West Germany had been rewarded for the statements by Moro in the Italian Senate and the letter from Scheel to Sadat in which the West German had declared his endorsement of the Arab interpretation of resolution 242.[18] It was suggested in certain Arab capitals that The Hague should follow these examples.

But the information from the Arab countries also seemed to indicate that matters were becoming more complicated. It was not so much the more radical countries that were now advocating a continuation of the embargo but more especially those that had hitherto been more moderate. Ambassador Van Hoeve reported Syrian sources as saying that Saudi Arabia was most strongly in favour of maintaining the embargo. The Ambassador in Algiers reported that Minister Abdessalam had told Jobert that it was not so much his country as Saudi Arabia and Kuwait that were most fiercely opposed to lifting the embargo against The Netherlands.[19]

At this time, relations within the EC began to change. Paris began to give The Netherlands more explicit support. Jobert gave the instruction to issue démarches in all the Arab capitals to get the continuation of the action against The Netherlands raised on the agenda. Van der Stoel regarded this in a positive light, though he could not avoid the suspicion that French self-interest also played some part. After all, the continuation of this discriminatory action made the start of a European-Arab dialogue, which would be so convenient to the French, scarcely thinkable. During a visit to Algeria, Jobert had insistently raised the issue of the embargo against The Netherlands and had even gone so far as to call it 'absurd'. Abdessalam's reply was that the USA had done far more to accommodate the Arab position. The Hague had offered no further comment on the November 6 declaration, as Van Elslande, Moro and Scheel had all done. When Jobert then answered that this hardly seemed a sufficient explanation, Abdessalam had said: 'You will find no objection from the Algerian side' and advised him to raise the issue in Jeddah and Kuwait.[20] All of which confirmed earlier reports from the Dutch embassy in Algiers.

On March 25, news came from De Ranitz in Paris that most Arab countries, according to *Quai d'Orsay*, seemed to have no objection to lifting the embargo. Even Damascus had by now come round to a more charitable position.[21] Only the Saudi Foreign Minister, Omar Saqqaf, had expressed hostility toward The Netherlands in Vienna, although he had promised to take up the question with King Feisal. This negative attitude, according to Dutch Ambassador Derksen, was in any case the result of the tactless behaviour of the American Ambassador, J. Akins.[22]

Meanwhile in Kuwait, '*Kiele Kiele Kuwait*' (Tickle Tickle Kuwait), the carnival hit from the Dutch makers of the TV satirical programme *Farce Majeure,* aroused enormous displeasure, with Rabbani being recalled to Kuwait. When it was revealed in the press there that the Dutch honorary consul had had connections with *Farce Majeure*, this caused a new storm of protest. In Kuwait the affair was taken up at a high level, as was confirmed in 1997 by the European commissioner F. Andriessen. When the French Ambassador had suggested lifting the embargo, the reply of Rashid-al-Rashid, the Secretary-General of the Kuwaiti Foreign Ministry, was to put this gramophone record on the record player.[23]

Plans for a United Nations Conference

Prior to the conference in Washington, there had been an interesting debate in The Hague. As we have seen, Jobert had strong objections to the basic premises of the Washington Conference. This was hardly surprising, since the American initiative was aimed directly against the essence of French policy vis-à-vis the oil crisis. As an alternative to talks between consumers, the French government proposed raising the whole issue of the oil problem at the UN, the most suitable framework being the Economic and Social Council (ECOSOC). Such an approach would from the beginning involve both producer countries and developing countries, a proposal that appealed to various members of the Den Uyl Cabinet.[24]

Although the aim of this plan was to accommodate the oil producers, representatives of the Arab and other oil-producing countries reacted rather critically to the proposal. The developing countries under the umbrella of the Group of 77 similarly were not impressed. From New York, the Dutch Permanent Representative, Fack, reported that the Egyptian mission thought the energy issue could not be raised at the UN in isolation, but should rather be investigated in the context of price trends among other raw materials and industrial products over the past 25 years. Besides, any treatment of the energy question would also have to include the role of the major oil companies. And furthermore, attention must also be given to the connection with world food prices and the high prices that developing countries had to pay for the transfer of industrial technology. Even Algeria expressed opposition to the French proposal. What was referred to in Western circles as the 'energy question' was, in the Algerian view, merely a part of the global issue of resources and development.[25]

Nor could the French proposal count on clear-cut approval within the EC. Most member states reacted with reserve whilst expressing regret that the plan had been launched without prior EC consultations. Italy did, however, adopt a cautiously positive attitude. The most powerful criticism came from the British side. The British Permanent Representative warned of the danger that the agenda of any energy conference might become so voluminous as to be unmanageable if developing countries all wanted to include their various resources.[26]

On January 22 Van der Stoel set out his stand against the French proposal. Whilst he did think that in the short term talks between consumers and producers were to be recommended – and he would bring this up at the forthcoming Washington Conference – he thought the UN conference advocated by the French demanded too much preparation time and, moreover, would lead inevitably to the vaguest kind of outcome simply because of the large numbers of participants. He had no principle objection to a global conference, but he would certainly prefer tripartite talks as envisaged in the follow-up to Washington.[27] It seemed a better idea, he thought, to leave the initiative in this to a leading developing country.[28]

Despite this reception, the French proposal did touch on an important – and in The Hague's view, a rather dubious – aspect of the Washington Conference, viz., the lack of consultation with producer and developing countries. The Dutch Cabinet had from the outset adopted the standpoint that such talks had to be started as soon as possible. The French proposal for talks within a UN framework went even further, however, than calling for the involvement of developing countries: in the UN, specifically in the General Assembly, the Western countries had no built-in majority. Quite the contrary in fact: nowhere was the influence of the Third World countries in principle so great as it was in the General Assembly.

There were different attitudes to the French proposal within the Dutch Cabinet, as in the earlier reception of Nixon's invitation. In a public address on January 22, Pronk came out in support of the French call for global talks; an address which caught the attention of the French Ambassador in The Hague, J. Senard. On January 24 Senard met Pronk for discussion, a meeting at which the latter, following the Foreign Affairs line, showed himself rather cool on some aspects of the French proposals. But Pronk recognised that what should be aimed at was a situation in which all concerned should be part of the consultative process. The Dutch Minister finally pointed out that the American proposal also talked about a global energy conference. In his view, such global discussion required a wider framework in which scarce resources other than oil should also be included for discussion. Nor should it be restricted to seeking agreements on raw materials but it rather should deal with prices, supply and demand.[29] In effect, he was allying himself with the standpoint of the more radical OAPEC countries like Algeria.

Senard succeeded insofar as Pronk was prepared to contend for the French proposal in the Council of Ministers of February 1. Pronk acknowledged that in several respects Jobert's way of conducting policy had to be repudiated: in particular, the French attempts to secure bilateral agreements. Whilst he could well see that for tactical reasons it was not possible to support Paris openly, he nonetheless thought that, from the point of view of the interests of developing countries, the French plan itself was one that he would subscribe to. Van der Stoel disagreed with Pronk; he considered Jobert's plan to be essentially an attempt to scupper the planned conference in Washington. Den Uyl, however, endorsed Pronk's judgement. However ambivalent the French proposal might be, he believed the oil problem had to be tackled in a global perspective. The Council in the end came to the compromise that the developing countries had to be involved in international discussions over the oil crisis as soon as possible.[30]

Aside from these deliberations, Permanent Representative Fack reported from New York that the French mission also realised that preparations for such a conference would cost time and effort. According to the

French there were two major problems. The first problem was the American attitude. Kissinger rejected the French proposal as premature. The planned consumer talks should first be held in Washington and then one should look further ahead. Secondly, the attitude of the producer countries was also less than encouraging. Jobert therefore had throttled back, though there were still considerable doubts in Paris over the usefulness of the rapidly approaching Washington Conference.

At the beginning of February, Algeria took up the French initiative by calling for a special session of the UN General Assembly dedicated to the problem of resources and development. The Dutch Cabinet had far fewer difficulties with this plan than with the French proposal. On February 7 Van der Stoel let it be known that he had no objection to this proposal, since the General Assembly session called for by Algeria would have a far broader character and would not interfere with the upcoming Washington Conference. At this stage, partly because of The Hague's positive assessment of the Algerian proposal, there began a certain Algerian-Dutch rapprochement that contributed to the shift in the Arab camp alluded to in the previous section. Some of the initially radical countries, particularly Algeria, gradually began to adopt the standpoint that it was now time that the embargo against The Netherlands be lifted.

This did not mean that The Netherlands and Algeria found themselves agreeing about everything, as was evident during a conversation between Dutch Director-General for Political Affairs Van Lynden and the Algerian Ambassador Chaalal. On that occasion, Chaalal gave his assessment of Kissinger's initiative as:

> an attempt to regain influence partly lost in the oil-producing countries and thus to re-establish [American] hegemony in the field of global energy provision both in relation to producers and Western Europe.

Therefore, Algeria advocated dealing with the whole issue in the UN. Van Lynden skirted round differences of opinion, pointing out to the Ambassador that The Hague's standpoint of working for a wider consultative framework after the conference fitted in with Algeria's wishes.[31] The same day Chaalal spoke with Pronk, who expressed his understanding for the Algerian criticism of the Washington Conference but at the same time asked that Kissinger's recent statements, to the effect that the aim of the conference was not to form a front of industrialised countries, should also be given attention. Pronk repeated that The Hague's standpoint on this matter in every way accommodated the Algerian objections.[32] On

February 13, The Netherlands and the other EC countries voted for the convening of a special session of the General Assembly.

The Sequel to Washington

As we saw in the previous chapter, it was decided at the Washington Conference to set up a high-level group – the Energy Co-ordinating Group (ECG), whose job it was to develop the implications of decisions taken. The ECG was also assigned to work on an *Oil Sharing Plan* to share the discomfort as widely as possible in the event of any new crisis.[33] Twelve countries had pledged to take part in the ECG: only France had declined any cooperation. As early as February 20, The Hague was invited for the first meeting of the group. The Americans were in a hurry.

At first sight, the Dutch Foreign Ministry thought Kissinger was moving too far ahead too fast, for a follow-up conference was to take place immediately after the special session of the General Assembly, now fixed for April. In Van der Stoel's view, there would for this reason be no interest in this follow-up conference, at least in the short term, among the developing countries. After the Cabinet session in which the invitation had been discussed, the acting Premier Van Agt remarked at a press conference that The Hague hated the thought of an excess of institutionalisation, and that the Dutch representative in the ECG should adopt a 'cautious approach'.

The high-level group met for the first time at the end of February, The Netherlands being represented by the Head of the Department for Economic Cooperation from the Foreign Ministry. During this meeting, which was mainly procedural, it became clear that the Americans had in the meantime abandoned their haste: the workgroups to be formed had until the end of May to produce their reports. At a second meeting eight workgroups were established, each to concern itself with a different aspect of the programme for international cooperation decided in Washington. To the satisfaction of The Hague, the workgroup concerned with the allocation of oil in time of crisis was assigned a Dutch chairman. The Dutch, after all, had for some time expressed great interest, especially within the OECD, in the creation of just such a crisis-allocation system in which the USA should also be a participant.[34]

Otherwise, however, the American policy toward the crisis did not give rise to unalloyed satisfaction. Washington seemed to adopt a highly aloof attitude to the preparation for a special session of the General Assembly, generating the impression that this session was somehow in conflict with the aims of the programme agreed in Washington. Such a conflict between Western interests and those of the Third World, as we have seen, had always been a cause for concern in the Dutch Cabinet. On March 16 Van der Stoel communicated his unease to the American State Department. He argued for a positive approach toward the aspirations of the Group of 77. He also wondered whether it would not be possible, as in Washington, to arrange for 'convergence' between the follow-up to the Washington Conference and the special session. The Netherlands further argued for such convergence at the third meeting of the ECG. Most other countries, however, wanted the special session and subsequent talks between producers and consumers to be kept separate. There was generally little feeling of support for any substantial energy discussion during the special session.

The Euro-Arab Dialogue

Jobert, meanwhile, was continuing with his attempts to get a Euro-Arab dialogue off the ground. Following the Washington Conference, the Dutch were anxious to give this dialogue a less anti-American character; for which purpose Van der Stoel, who was greatly disturbed by the conflicts within the Atlantic world, travelled to both London and Bonn for talks. On the advice of his Director-General for Political Affairs, Van Lynden, and against all his own objections, he decided to go along with the announcement of the EPC plans for a dialogue, on the condition that only a first, exploratory, stage should be initiated. There could be no question of second and third stages, involving the setting up of a mixed commission and a European-Arab ministerial conference, as long as the embargo against The Netherlands remained in force.[35]

During an EPC meeting in Brussels on March 4, the Nine formally accepted a French initiative to pursue such a dialogue. Only the British were unwilling to bind themselves because of the change of government in London where the Labour Party under Harold Wilson had just been elected to power. Van der Stoel later said that he had only agreed to the first exploratory stage of this dialogue because

> the Nine as such were incapable of moving on to the second stage, let alone the third, so long as the Arabs governments of their own choice continued to wield the weapon of discrimination against the group of Nine.

Furthermore, the dialogue must not be allowed to undermine the ECG consultations or the American peace efforts in the Middle East.[36]

It is unclear whether the eventual decision issuing from the EPC deliberations was to go for the first stage of the Euro-Arab dialogue or the dialogue in general, albeit in that case with the Dutch conditions attached. In the Dutch Council of Ministers, Van der Stoel informed his colleagues that he had gone along with the dialogue proposal in the EPC to avoid being isolated, but that two conditions had been stipulated. The first was that the existing peace negotiations must not be hindered; the second, that following the second stage there must be renewed talks over further progress.[37] In the First Chamber he mentioned yet another condition: that the energy conference must not be thwarted.[38]

On further reflection, it had been a remarkable meeting in Brussels. At that same moment Kissinger was visiting NATO headquarters, but there was no contact with the EC ministers even though they were only a fifteen minute journey apart. Kissinger was not informed of the decisions made in the context of European Political Cooperation. On his return to the USA, Nixon sent a tough letter to Chancellor Willy Brandt in which the Euro-Arab dialogue was rejected in some fairly sharp language.[39]

No consultation had taken place. It is not impossible, however, that during the EPC discussions Scheel did suggest having had a meeting with Kissinger. He even gave the impression that the previous day the Americans had agreed to the dialogue idea, which in turn had been a reason for the Dutch readiness to compromise. Van der Stoel was therefore unpleasantly surprised when it became apparent that the State Department was reacting negatively to the Nine's decision to arrange a dialogue. Both in Cabinet and in the Second Chamber, considerable resentment was expressed at this course of events. Brandt later attempted to soothe matters by assuring Washington that the Nine had taken into account the American objections to the dialogue being set in motion too soon.[40]

This could not have been easy, because the French pursuit of a Euro-Arab dialogue was disparaged in Washington in extraordinarily negative, dismissive terms. 'The idea of a Foreign Ministers' meeting between all Arab states and European states can only fill one with horror,' said Kissinger at a staff meeting at the State Department. 'Anyone seriously pursuing it ought to have his head examined.' One of the most important American objections was that a Euro-Arab dialogue under French leadership would strengthen Algeria's position at the expense of the Egyptian Sadat, who was highly regarded in Washington.[41]

On March 12 the *Comité Politique* met to discuss this Euro-Arab dialogue further, agreeing, at least after British consent, that the Arab countries should be informed immediately of the main lines of the decision taken. At once, the old differences on the dialogue's basic aims and principles reasserted themselves, this time the question of how far the dialogue should be concerned with the Middle East peace settlement and related political questions. The French stressed that if economic cooperation were going to be discussed with Arab countries, it was simply unrealistic to exclude political questions from consideration. The Dutch delegation warned again of the danger of undermining other initiatives. According to The Hague, the aims of the dialogue had to be economic, not political. There was also disagreement over the Israeli position in the dialogue. It was agreed that there had to be contact with Israel; but the French wanted this to be held in strict secrecy in order not to prejudice the dialogue with the Arab countries. The Netherlands and Denmark expressed their dissent: the Nine had to maintain a certain balance in their readiness to cooperate with countries in the Middle East. And, moreover, since cooperation with Israel was already on the agenda, this could not now be ignored.[42]

However, as we have seen, all this had to be shelved until it was clear what the new British Labour government's attitude to the EPC plans would be. This exchange of power in London was highly significant for the Dutch Cabinet. Labour, like the PvdA in the Netherlands, had neither an anti-American nor anti-Israeli reputation. Wilson's government might therefore mean significant support for the Dutch; and Van der Stoel lost no time in travelling to London, in early March, to hold discussions with his new colleague, James Callaghan. Callaghan, it turned out, backed the dialogue. He had reservations, though, being particularly anxious not to accentuate differences with France. In general, the new party in power in London clearly gave priority to cooperation with the USA, more so than the previous Conservative government under Heath. Van der Stoel, of course, had been playing precisely this tune for some time, and so could speak freely of the desirability of more frequent consultations with the USA.[43]

It had also not been lost on Kissinger that the new Labour government advocated a far more pro-Atlantic policy than Heath's government. During talks with the representatives of the American *majors*, the American Secretary of State declared that 'international affairs have reached a point where a Labour government is more pro-American than a Conservative one'.[44] Kissinger could draw a similar conclusion with regard to the Dutch government of Den Uyl and Van der Stoel.

When it became clear on March 18 that the embargo was not to be lifted, The Hague's resistance to the Euro-Arab dialogue increased. Immediately following the OAPEC decision, Van der Stoel reminded the EPC chairman Scheel of the qualification he had entered; the second stage of direct official contacts with the Arab countries could not begin as long as the embargo remained in place. This reminder was also brought to the attention of all EC governments. For Paris, this block on further progress that seemed to have arisen was all the more reason to call for the Arab world to lift the embargo against The Netherlands.

All the problems surrounding the Euro-Arab dialogue once more illustrated how sharp the differences were between the EC member states. In The Hague, Van der Stoel submitted a note on these problems on March 21 which was discussed on March 27 in the Council for European Affairs and on March 29 in the Council of Ministers. Van der Stoel's note contained the classic argument for Atlantic unity with the corresponding view of European integration: distrusting European Political Cooperation and advocating strengthening of the Commission. The note suggested that cooperation within the EC was threatened with paralysis, at least partly as a consequence of the unwillingness to strengthen the authority of the European Commission. The energy crisis and the admission of three new members had further undermined the willingness to act, and the more prominent role assumed by EPC had put the European-American relationship under some pressure. The note therefore gave the highest priority to combating the acute crisis of confidence that had arisen within the relationship between the Nine and the USA. It was most important to improve these relations. If Paris was unwilling to cooperate, then one must turn to Bonn and London.[45]

At the end of March, there was a positive response from the Arab world to the proposal from the Nine to institute a dialogue. On April 20 and 21 the nine EPC Ministers gathered informally at Gymnich Castle in Bonn to discuss this response. Once again, there were differences of opinion. The first of these centred on the Dutch pre-condition. Jobert refused to commit himself in writing to the principle that the discriminatory Arab actions against certain EC countries must first be repealed before the second phase of dialogue could begin. Van der Stoel accepted this, under the proviso that the Nine did consider themselves bound to the spirit of this qualification. This was a proviso his eight colleagues were prepared to accept.

A second problem concerned relations with Washington. Jobert, perhaps afraid of 'renewed treachery' by Scheel, turned against the proposal that the EPC chairman should be entrusted with full and prompt exchanges of information with the USA during the dialogue. Despite French opposition, Scheel pushed through the so-called 'Gymnich formula' which 'gave the USA the right to be consulted in the EPC decision-making process'.[46] During this EPC gathering, Jobert, according to Van der Stoel, had adopted an 'extremely anti-American standpoint'.[47] Van der Stoel kept quiet on this question to avoid exposing himself unnecessarily, and agreement was thus eventually achieved.[48]

These differences of opinion again demonstrated that the Nine could only reach any kind of agreement on the dialogue with some difficulty. The pursuit of a European-Arab rapprochement therefore arrived at a complete impasse when its driving force, Georges Pompidou, died on April 2. With the death of Pompidou, after Heath the second greatest champion of an independent European role on the world stage had disappeared. In May, Valéry Giscard d'Estaing was elected President, initiating a period during which French foreign policy would be much more Atlantic-oriented.[49] Shortly after this, the curtain also fell on Willy Brandt's time as Chancellor. Brandt's position had been steadily weakened during the oil crisis. Finance Minister Helmuth Schmidt had drawn power to himself, a fact that had become increasingly obvious during EC meetings.[50] It was therefore not by chance that following Brandt's resignation in the wake of a spy scandal, he should be succeeded by the more pro-Atlantic Schmidt. In the spring of 1974, politicians thus came to power in three major EC countries who were far more Atlantic-oriented than their respective predecessors.

The Supply Recovers

As we saw earlier, the oil supply to The Netherlands had by now begun to recover. It had already become clear during the period of rationing that the worst was over. In mid-January the Dutch Cabinet still believed that the supply would be 70% of the previous year, whereas the actual supply turned out at the end of the month to be over 80%.[51] Since processing and consumption were lower than the previous year, the stocks at the end of January were 20% higher. The situation was roughly similar in February and March: supply stabilised at around 80 to 85% and processing and consumption recovered to around 85%. Whether the reduced supply was still a consequence of the embargo was a question the Cabinet doubted. The Minister for Housing, J. Gruijters, asked for clarification in the

Council of Ministers of March 8. According to Lubbers, the reduction was indeed the result of the embargo.[52] In March, however, stocks reached a level that was in comparison with the previous year as much as 30% higher. Throughout the whole embargo, the end stocks of crude oil remained higher than the previous year.[53] The situation was even more reassuring in April, with supplies more or less reaching the level of the previous year. Processing recovered to almost 95% of the level for 1973. Meanwhile, Shell communicated that there were no more immediate problems as far as provisioning was concerned.[54]

After the period of rationing, the need to restrict the consumption of oil products was thought, logically, to become gradually less. In a letter to the Second Chamber, Westerterp announced on February 1 that maximum and minimum speed limits on the roads would be introduced from February 6. Moreover, the action *3 x 5 = 15* was started by the Ministry of Transport and Water Management together with the Dutch motorists association and the Association of Automobile Dealers and Garage Owners. There would be a 5% saving from the introduction of the maximum speed limit, 5% from a more economical style of driving, and 5% from more selective use of motor vehicles.[55] The public readiness to observe these prescriptions, however, was not what it had been at the beginning of the crisis. This was hardly surprising, given the steadily improving reports of the oil supply. Measures restricting consumption, such as cutting back on the use of electricity for street lighting and advertising, were in fact also lifted several weeks later.

'Sweating it out'

Yet the embargo, at least formally, was still in place. At the Foreign Ministry toward the end of March, the balance was again calculated. This was partly as a result of reports such as that Scheel had sent a grovelling letter to Yamani and Abdellasam. Parts of Scheel's letter had by then appeared in the Algerian press. These reports increased the pressure from his civil servants for Van der Stoel also to go further to meet the Arabs.

But what exactly had Scheel written? Every attempt was made to get hold of the full text of the letter. On March 28, De Beus reported that the missive contained a record of the German explanation of their position during the meeting with Yamani and Abdessalam. Subsequently, the Arab side had requested a written statement of the German standpoint, while in Bonn no objection had been raised. De Beus had pointed out to his interlocutor from the West German Foreign Ministry that the wording was rather pointed, in particular the reference to returning to the pre-1967 borders and the demand for complete withdrawal of foreign troops. De Beus had also requested a copy of the letter, a request which had caused a certain German embarrassment.[56]

The result of this letter in The Hague was to increase the pressure on Van der Stoel. On March 28 he had a meeting with the Algerian Ambassador. In the preparations for this meeting, Director-General Van Lynden hinted that Scheel's letter should provoke thought. He suggested that Van der Stoel too 'could write something similar'. 'I know that the idea does not appeal to you,' acknowledged Van Lynden, 'but as a result of their interpretative statements or their letters, your Belgian, Italian and German colleagues have put the Dutch Government in a dilemma.'

Van der Stoel again put the suggestion to one side. 'I fear', he replied, 'that in our case this would lead to further demands for clarification (for example, 'also Jerusalem').' In fact, this very question of a clarifying statement was raised during the meeting with the Algerian Ambassador, who took the opportunity to suggest that Algeria could publish an interpretation of the Dutch standpoint that would not subsequently be denied by The Hague. Van der Stoel, however, did not take up the offer. He replied that his standpoint over the Israeli withdrawal was no different from that of the other EC partners.[57]

Partly as a result of this, the Head of the Department for Africa and the Middle East (DAM) drafted a note in which The Hague's position was analysed further. Three options were identified. The first was the suggestion of the Algerian Ambassador, that Algiers should publish an interpretative statement of the Dutch position. This should state that the EC formulation of November 6, '*mettre fin à l'occupation territoriale*', meant in The Hague's view, '*evacuation de tous les territoires occupés*' (i.e. *ending* the territorial occupation meant *complete withdrawal from* all occupied territory). The second option was for the Dutch government to issue its own clarification of the EC November 6 statement; and the third option was '*sweating it out*'.

The Algerian suggestion undoubtedly had its advantages, concluded the Head of DAM. The Hague need not make a further statement itself and would yet satisfy those in the Arab world who no longer wished to maintain the embargo weapon. Should this option be taken up, it must be established in advance that both formally and in practice the embargo would be lifted by all Arab countries and that there would be no propagandist use made of it (e.g. 'Netherlands caves in'). There should further be plenty of opportunity provided for resumption of trade with the Arab world, and the Arabs must cease working up cultural indignation over such trivial issues as the *Farce Majeure* song, 'Kiele, Kiele, Kuwait'.

On the possibility of issuing a Dutch statement, the DAM note suggested that the best approach would be to follow the West German example, to publish a summary of discussions held. The talks Van der Stoel would hold outside the meeting of the General Assembly would provide a good opportunity. The memorandum came with an attached draft of such a statement, setting out that a peace settlement must be found that could be agreed on by all parties, which incorporated coherently all the elements of resolution 242, as repeated in the Nine's declaration of November 6, 1973.

> In accord with this resolution, the borders existing before 1967 should be taken as a basic principle.

This implied

> ending the foreign territorial occupation of national territory and thus a complete withdrawal of foreign troops. Border corrections, reached through peaceful talks and willingly effected by the parties concerned, should not of course be excluded.

Something should be said in a possible final paragraph over 'the importance of American attempts to bring the respective parties together'.

The third option, finally, meant maintaining The Netherlands 'principled position', i.e. waiting until the embargo was lifted through the play of 'exogenous factors'. An end to the embargo might finally result from various factors: pressure from the Nine; as a lever to get a stalled dialogue kick-started again; or through the growth of awareness within OAPEC that continuation of the embargo promised less advantage than disadvantage to the Arabs; or possibly American pressure – 'which we still hope for'. It was apparently assumed at the Dutch Foreign Ministry that the USA was not at that moment about to exert any pressure on the Dutch behalf.

It was clear from a note added by Van Lynden that he was a strong advocate of the Dutch issuing their own statement.

> Now that the Belgian, Italian, and German statements have given their own interpretations to the November text, the Dutch government, in my view, should not avoid giving its own interpretation.

The appended text was in his view considerably less drastic a revision than the Italian text and slightly less than the Belgian and German statements. According to Van Lynden, putting out an interpretative statement, either now or at the latest in New York, would have the best chance together with the French attempts of succeeding in getting the embargo lifted.

Yet once again Van der Stoel flatly rejected the advice of his Director-General for Political Affairs. This was made very clear in a note that he added to the DAM memorandum: 'Choose option III; text appended to option II too radical for me and moreover would lead to further demands for interpretation', and that was an end to the matter.[58]

In Cabinet, too, Van der Stoel adopted the same unbending stance. He urged his colleagues not to be too gloomy. The Arab countries were beginning to see that continuing the embargo would only impede Euro-Arab dialogue. Den Uyl shared this point of view and criticized Scheel's action. The fact that news had only been received of his letter weeks later was a matter that should be taken up at a high level. A protest was in order. Lubbers, as on earlier occasions, was less optimistic, pointing out all the negative consequences of the position to which The Netherlands had been brought. Transport Minister Westerterp added his own comment, too: it must remain the aim of Dutch policy to get the embargo lifted as soon as possible; and in this connection, he pointed out the problems under which KLM was still labouring.

Van der Stoel recognised that the embargo was a serious headache, but repeated that if one were to go down the same path as Scheel, new Arab demands would surely follow. He proposed getting the EC partners once more to exert pressure in the Arab capitals to get the embargo lifted. A similar request should also be made to the American government, and beyond that, bilateral goodwill initiatives might be considered. It was eventually decided that Van der Stoel should protest against the steps taken by the Italian and West German governments. Secondly, the EC partners would be asked to demonstrate solidarity and not to accept discrimination against a fellow member state.[59]

On March 29 Van der Stoel took a far more lenient view of the West German letter. He concluded that Scheel had refused an 'additional statement' and had only given a written account of the conversation with Yamani and Abdessalam. The earlier conclusion, that The Hague had refused to provide an additional clarification to the EC declaration of November 6 whereas Bonn had complied, was accordingly incorrect.[60]

The Special General Assembly

The special session of the United Nations General Assembly so desired by the developing countries was held in April. Initially, the American government was against such a session, which was considered a hindrance to the consultative process resulting from the Washington Conference. But partly due to pressure from the EC countries, the Americans eventually came round to participating in the special session that opened on April 9 and lasted for three weeks.

The view of this session taken in The Hague was not entirely positive. As we saw, Lubbers in particular had expressed his doubts in the Council of Ministers. His view was shared at the Ministry of Economic Affairs, particularly within the Directorate-General for Foreign Economic Relations. In a note from the Director-General, the special session was situated in the context of an increasing loss of Western influence in the world. Up till then, international economic cooperation had been based on GATT and the IMF. The basic principles grounding this international order had been undermined, according to the note. There were two reasons for this. On the one hand, there had been increasing economic nationalism and a concomitant lack of adequate international co-ordination; and on the other hand, there was the increasing influence of power formations in economic life. In addition, the power that the Bretton Woods system had always carried was being eroded. The United States was still trying to take up the leader's role, but this was being thwarted by Paris as well as several developing countries. The position of OPEC was largely so strong because of Western divisions, and OPEC was always ready and able, if not always liberally, to act as the standard bearer of the interests of the developing world.

All these developments were accelerated by the oil crisis. 'If my view is correct, we are now living through the last days of Bretton Woods and of GATT', wrote the Director-General. The Western countries were in a weak position vis-à-vis the developing countries because of the dependence of their economies and their mutual divisions. Some countries were trying to secure their own interests by means of bilateral agreements, leading to even greater confusion. It was significant that a new Charter was already being discussed within the UN, one that would no longer be based on the principle of free trade. The concluding advice of the note from the Director-General for Foreign Economic Relations ran:

> We have to try to provide a realistic counterbalance to the rather grandiose philosophies of the Ministry for Foreign Affairs that would seem to betoken a flight from reality.[61]

It is certainly true that the special session was regarded at the Dutch Foreign Ministry in a much more positive light. Such views were in the first place inspired by the progressive climate of ideas within Pronk's Directorate-General for International Cooperation; but neither the Directorate-General for Political Affairs nor Van der Stoel was against the idea of the special session, albeit for more opportunistic reasons. For them, the session seemed a good opportunity for improving relations with the Arab countries.

The special session kicked off in New York on April 9.[62] The following day, the general deliberations were opened with an address by the Algerian President Boumédienne, who declared that the non-aligned countries were now a new driving force in international relations. The non-aligned countries sought to establish fair participation for the Third World in global politics, which had to be based on a fair share of the world's wealth. This necessarily required a fundamental reorganisation of economic relations between rich and poor countries that would lead to a redistribution of the possibilities for growth and development.

Over the course of the following days, the Western countries, including specifically the EC member states, acknowledged that a fairer share in the world economy should be striven for. Such terms as 'collective economic security', 'economic peace-keeping' and even a 'global early warning system' (against undesirable fluctuations in the market for raw materials). Even the USA declared, out of Kissinger's own mouth, a willingness to contribute to the economic development of the Third World, including the transfer of technologies.

On April 17 the oil-producing countries reviewed the history of the oil price increases: a long overdue adjustment of price levels, according to them. They pointed out that the welfare of the industrialised countries had largely existed thanks to the fact that oil prices had been held at an artificially low level. In this regard, it was striking that the Saudi address was more moderate than the other oil producers. Yamani argued that polarisation had to be avoided. In his view, the danger of a global recession as the result of uncontrolled price increases was so great that this would in all probability lead to a 'devastating global war'. Because of the developed economies' greater capacity for adapting, the developing countries would be the losers.

Van der Stoel, in his address, adopted a positive attitude toward the wishes expressed by various Third World countries. He stressed that from the beginning The Hague had tried to play a mediating role. The Netherlands hoped that a clash between North and South could be avoided and that the deployment of a broad plan of action would be adopted with the general support of the industrialised world. He appreciated the work of the Group of 77 and expressed a willingness to work with the Group toward the realisation of a new international economic order.

Partly at the urging of the Dutch, the Nine declared their readiness to contribute to a fund for the worst hit developing countries. On behalf of the EC, Scheel advised Waldheim on April 26 of an offer of a 'substantial contribution'. The Dutch delegation was positive over the Algerian President's proposal to set up short-term funds. If this should not prove feasible, support would be given to set up a preparatory committee whose task would be to submit concrete proposals to the UN Economic and Social Council within a short period. The proposal for a *'crash programme'* for 1974, ahead of a fund for the worst hit countries, was also actively supported by The Hague. And finally, The Netherlands gave its support to the proposal to set up a consortium on behalf of those countries most affected.

With the winding up of the general deliberations on April 24, the real work began behind the scenes. In the meantime, a workgroup had begun on April 11 with the framing of a development statement and a draft action programme. Away from the floodlight of publicity, the Western countries took a rather less cooperative line than they had professed during the general deliberations of the General Assembly. The traditional oppositions between developed and developing countries re-emerged during these negotiations. It was also striking that whereas the energy question and oil prices had assumed a prominent place in the general deliberations, the negotiations focused on the more *general* questions of development, raw materials, trade and finance. This was mainly because the Group of 77 rejected every reference to the energy question. A French attempt to deal with the oil problem within the wider context of the question of raw materials in a committee of producers and consumers was wrecked by the resistance of both the producers and the USA.

Albeit with the greatest difficulty, some sort of agreement was reached in the workgroup on the framing of a draft statement and an action plan; but this could not disguise the fact that in several respects the opposition between North and South remained insurmountable. At the final session of May 1 and 2, several countries made a final statement. The American

Permanent Representative stressed that it would be incorrect to speak of consensus, given the number of points on which there was no unanimity, particularly referring to the action plan. The USA had objections to this programme without going into details. Other Western countries said they also had important reservations.

On the other hand, the Algerian Foreign Minister, A. Bouteflika, spoke of a consensus that had no precedent. He stressed that this did not mean unanimity, but added that unanimity should not be expected if one were dealing with a fundamental change in the economic system. The session had strengthened the UN, to his mind, while the conduct of the Group of 77 had been marked by a spirit of moderation. He pointed out that the General Assembly had witnessed a change in power relations although, in the light of the attitude of most Western countries, it is highly doubtful whether Bouteflika was right in this.

An important motive for Van der Stoel's trip to New York, as mentioned earlier, was to try to make contact in the corridors with colleagues from other countries, not least the Arab countries. Van der Stoel thus held conversations with counterparts from Egypt, Kuwait and the United Arab Emirates, among others, all three of whom were for the lifting of the embargo. His Egyptian colleague assured him that no bilateral problem existed between Egypt and The Netherlands, and that he would present the case to his government for lifting the embargo. This sounded altogether positive. Nevertheless, the Egyptian Minister emphasized that The Hague would have to come up with 'something new' before the OAPEC meeting in June.[63]

Later in Washington, Van der Stoel also held talks with Kissinger, pointing out to him the remarkable fact that Saudi Arabia especially refused to cooperate in getting the embargo lifted. Kissinger promised to raise the question on his following visit to King Feisal. The two statesmen differed, however, in their estimates of the value of the special session. Van der Stoel thought the session far less confrontational than he had anticipated and wondered whether agreement might be possible, in the sense that a UN world conference on energy might be arranged, attended by a representative delegation from both producer and consumer countries. Kissinger's response was negative. He reiterated the familiar American standpoints: that better cooperation had first to be achieved between the consumer countries; that a UN energy conference could have disastrous consequences if no prior agreement had been reached between the Western countries. Kissinger thought it absurd that the producers should be capable of forming a cartel whilst the consumer countries could not.[64]

Subsequently, Van der Stoel concluded that hardly any progress had been made at the special session. There was possibly the draft of an action plan, but that was virtually crippled by a series of preconditions. In the General Assembly, the initially positive tone had degenerated into confrontation.[65] In The Hague, nonetheless, the session was not seen as a disaster: indeed, far from it as far as The Netherlands' status was concerned. For the attention had shifted during the session away from the politics of energy to the more general struggle of the Group of 77 toward a New International Economic Order. Most Western countries had reacted to this demand by rejecting it outright. The Dutch delegation, however, had adopted a more sympathetic attitude and within the EC had endeavoured to foster understanding for the proposals of the Group of 77, an attitude which, according to Van der Stoel, had been highly appreciated by various developing countries. It was therefore concluded in The Hague that adopting this approach had certainly contributed to the view now taken by a number of Arab countries that the embargo should be lifted.[66]

Saudi Arabia Stands Firm

While Van der Stoel was busy in New York, a certain optimism briefly prevailed in The Hague concerning the likelihood of the embargo being lifted. This feeling arose through talks held between Den Uyl and the Saudi Ambassador Nowilaty. The Premier inferred from the latter's wording that there was now a great chance that the embargo would be lifted, and on April 11 he shared this judgement with the Council of Ministers,[67] going on to speak of his optimism at a press conference. This of course led to premature news reports that the end of the embargo was in sight, leaving Van der Stoel, on his return, to have to straighten various matters out. In the Second Chamber, Den Uyl later denied that he had ever uttered anything definite about the duration of the embargo, but the impression he created was less than highly convincing.[68]

Indeed, Den Uyl had been too optimistic: the embargo still had months to run. It was becoming clearer at this stage that especially the moderate countries, Saudi Arabia and Kuwait in particular, remained opposed to any lifting of the embargo. In both countries 'the familiar grievances' were still playing a role, but in the case of Saudi Arabia there was another line of reasoning involved. As early as March 28, Dutch Ambassador Derksen had reported that Jeddah was not, on the whole, inclined to lift the embargo, but rather saw it to their advantage to 'keep Western Europe divided'.[69] It was also clear from other sources that Saudi Arabia had no interest in promoting a Euro-Arab dialogue.[70] Maintaining the embargo therefore served the purpose of blocking this dialogue in a highly effective manner, an entirely different function from the early days.

In May the Dutch Cabinet adopted a more wait-and-see attitude. There was no doubt by now that Algeria wanted an end to the embargo. In Den Uyl's judgement, the Algerian standpoint was partly based on the attitude shown by The Netherlands at the special session. Furthermore, the Dutch Foreign Ministry was supporting the Algerian Minister Bouteflika's candidature for the chairmanship of the General Assembly, and at the same time attempting to find a mutually satisfactory compromise with Algeria in the World Bank.

At the end of May, Van der Stoel expressed the hope that the coming Israeli-Syrian disengagement accord would lead to the end of the embargo.[71] The accord was signed in Geneva on May 31, the same day that the Security Council approved a resolution empowering Waldheim to take whatever measures were needed to set up a corps of observers to supervise the disengagement due to begin on June 1.[72]

Van der Stoel considered taking a trip himself to several Arab capitals after these events,[73] but, to much surprise, no decision was taken at the OAPEC session on June 2 to lift the embargo against The Netherlands. By now, a complete reversal of positions within the OAPEC had taken place. On June 3 the Dutch Ambassador in Damascus reported an assurance he had been given by the Syrians that their country had not voted against lifting the embargo. On the question as to who had, the Syrians pointed out that 'there were Arab countries that regarded the Euro-Arab dialogue with great reluctance'. In diplomatic circles, Saudi Arabia, Kuwait and Qatar were now always cited as the most important opponents of lifting the embargo. On the other side, Algeria explicitly declared itself opposed to maintaining the embargo. In fact, Algiers let it be known that it was no longer willing to abide by the decision of June 2 to keep the embargo in place.[74] That other radical state, Iraq, had also spoken out against maintaining the embargo.

Various news reports reaching The Hague during June confirmed that Saudi Arabia in particular wanted to keep the embargo in place in order to obstruct any Euro-Arab dialogue. More unpalatably, Saudi Arabia was being supported in this policy by the United States, or was even acting at the behest of the USA. The Algerian Ambassador in East Berlin suggested that Jeddah had not been acting autonomously in the maintenance of the embargo, 'but rather in the present case had been little more than an instrument of the United States'. It was Washington that had got Jeddah to resist the lifting of the boycott

> because by this means the American government thought it could see a way of obstructing the dialogue and cooperation between the EC and the non-European Mediterranean states.

The Dutch Ambassador in East Berlin found this argument highly improbable, but the Algerian stuck to his story.[75]

Similar stories were coming from other sources, not least from Paris. De Ranitz reported that a highly placed official in the French Foreign Ministry, de Commines, had confirmed the rumours – also printed by *Le Monde* – of American duplicity. Kissinger, for various reasons, was against a Euro-Arab dialogue and had probably convinced King Feisal of the undesirability of such a dialogue. It was not inconceivable that he had told the king that only Washington was capable of forcing Israel to make concessions. Furthermore, only the US would have been capable of combating the communism feared so much by Feisal.[76]

Reports that Washington had incited Feisal to adopt a tough stance were also circulating in the Egyptian press. Although Van der Stoel subsequently claimed to have given these suggestions little credence, they were certainly believed at the Foreign Ministry in 1974.[77] In a memo to the Dutch Foreign Minister on June 7, the acting head of the Department for Africa and the Middle East concluded that it was doubtful whether it was of any further use, given this background, to continue urging Kissinger to put in a good word for The Netherlands.[78] On the same day, Van der Stoel referred in the Council of Ministers to the reasons why Saudi Arabia was opposed to a Euro-Arab dialogue. There was nothing in it for the Saudis, whereas Egypt and Algeria, their competitors, stood to take a more prominent role in the context of such a dialogue.[79] A week later, the Dutch Ambassador Derksen reported from Jeddah that both his French and British colleagues thought the USA and Saudi Arabia were in league in this affair. The Ambassador confessed that he 'had to acknowledge a certain reality in the scenario alluded to', not so much because of *Le Monde* or Israeli propaganda but because a highly placed Saudi official had confirmed the story's truth.[80]

The OAPEC decision of June 2 had been a blow. Several days later, Den Uyl expressed his concern over Rotterdam's position now that the embargo was to continue officially. On June 12 the Mayor and Aldermen of Rotterdam again asked the Premier as a matter of urgency, given the magnitude of the interests at stake, not to relax the pressure and 'to take whatever steps were necessary to get the Arab countries to end the boycott against The Netherlands'. Yet again, Van der Stoel decided to turn to Washington to mediate, but without result. According to Arab information, Feisal took a sharp line when Nixon paid a visit to Jeddah. Minister Saqqaf told a British diplomat, moreover, that any removal of the embargo in the short term was out of the question 'unless the Dutch government should publicly call for a total Israeli withdrawal from Arab territory'. At a ministerial session of NATO, Kissinger was also in a 'fairly hopeless' mood.[81]

One of Feisal's influential advisors suggested to the French Ambassador that the embargo would be withdrawn at the next OAPEC meeting on July 10. But on June 18 De Ranitz reported that King Feisal himself remained adamant.

> The matter was complicated by the rumours regarding the anti-Dutch intrigues of the United States which in all probability had taken root in Jeddah and were even being confirmed from the Saudi Arabian side.

The new French Foreign Minister, Jean Sauvagnargues, had also referred to these rumours.[82]

In London, suggestions of American double-dealing were being treated cautiously. The idea was not considered probable, although it could well be that the USA, given the undesirability of a Euro-Arab dialogue, was not exerting itself as strenuously as it might to get the lifting of the embargo discussed as a matter of urgency.[83] Yet the reports persisted. The Algerian Ambassador in Peking also confirmed this reading of the American role, adding that this attitude was mainly due to the influence exerted by the American-Arab oil company Aramco.[84]

A short time later, however, the Saudi Minister Saqqaf spoke of The Netherlands in a rather more conciliatory fashion. There were deliberations within the Dutch Foreign Ministry over the possibility of offering Saqqaf a 'face-saving' formula in the guise of a letter reiterating the Dutch standpoint, but Van der Stoel would not hear of this initiative. 'In the light of earlier experience with the Saudis on this point,' he said, 'I think it would be extremely dangerous.' It could create the impression that a price can still be demanded for the lifting of the embargo.[85]

In fact, it was decided at the OAPEC meeting on July 10 that the embargo against both Denmark and The Netherlands should be lifted. Little attention was paid to the matter within the Arab countries since for some time the embargo had been regarded as an affair superseded by other events, and moreover, it had for some time no longer been observed by many countries. By this time, most other anti-Dutch measures in Arab countries had also been rescinded. The boycott of KLM by Libya had been lifted at the end of April, though in Iraq this did not happen till August, and the boycott of Dutch shipping by dock-workers in Libya was not withdrawn until September.

On the day the embargo was lifted, NRC *Handelsblad* wrote that 'a political comic strip that has scarcely any comparison in the entire history of Dutch foreign policy' had now reached its end. Future historians would undoubtedly be poring over the days of '*grandeur et misère* behind the dykes'. Besides which, according to NRC *Handelsblad,* the Arab attitude had also been based on the knowledge that plans for a Euro-Arab dialogue could only be implemented if the embargo were lifted.[86]

The Energy Co-ordination Group

The Netherlands was of course much relieved at the lifting of the embargo. But in one respect The Hague could also be satisfied. A day before the end of the embargo, during a meeting in Brussels, the twelve countries that made up the Energy Co-ordinating Group (ECG) had decided 'in principle to pool member nations' oil resources in the event of a future energy crisis'.[87] The ECG talks had by that time got into their stride. On the basis of a Foreign Affairs note, this progress was discussed in the Council of Ministers in The Hague on June 14.

There were still several important problems on the horizon, said the note. There was little chance of talks between producers and consumers in the foreseeable future. During the special session, it had been evident that the time was not yet ripe for a special summit devoted to energy, involving producers, consumers and developing countries. There was little enthusiasm, especially in Washington and London. Nor did the producer countries seem very keen; they seemed rather to want to maintain a free hand. Within the ECG it was sometimes wondered whether there was any point in pursuing a dialogue with the producers, not least because by this time the oil prices had begun to fall again. From the Western point of view, such talks were no longer urgent, but for The Netherlands such a discussion was still desirable, at least if a clash with the producer countries could thereby be avoided.

Ideas on the role and status of the international oil companies had been widely discussed within the ECG. A report on this was drawn up, with the most important conclusion that the consumer lands had to form a closer, more collective relationship with these companies which, it was taken for granted, would continue to play a central role in the future. It was not the aim to interfere with the power of the oil companies. What was needed was to set up an effective international system of information.[88]

This view conformed entirely with the interests and standpoint that The Hague had always advanced throughout the oil crisis. The most important problem the Den Uyl Cabinet had had to wrestle with had been in dependence on the oil companies for information, a state of affairs that had bred insecurity and distrust. The delegation in the ECG, in the view of the Dutch Cabinet, should therefore urge that the oil companies be legally compelled to divulge information in times of crisis. This might have to happen through a mutual adjustment of the different national legislations. More generally, and preferably within the context of the OECD, a code of behaviour should be introduced that would hold not only in times of crisis.[89]

In mid-July it was evident that the ECG had largely succeeded in reaching an accord on an *Integrated Emergency Programme* that, inter alia, provided for the sharing of oil in cases of serious reductions in the oil supply. The programme was based on three basic principles: 1) participant countries, in proportion to their normal imports, must lay up a stock of oil for a number of days to be determined later; 2) they must jointly reduce consumption if the oil provision for the entire Group should fall below a certain percentage, viz. 7%; and 3) subject to decisions to the contrary, the participants commit themselves in such cases to switch to oil-sharing according to an automatic formula.

To an extent, this scheme corresponded with the allocation scheme already worked out within the OECD prior to the oil crisis. It was even more comprehensive, since the obligations to lay in stocks and to limit consumption in a crisis situation were now defined and incorporated into the allocation ratios. According to the Dutch Foreign Ministry, this was cause for satisfaction, because agreement now seemed possible for an *Emergency Programme* that would serve to anchor the solidarity of the participant countries. The Council of Ministers shared this positive assessment.

Finally, the ECG was to draft an International Energy Programme that would be based on the International Energy Agency (IEA) founded in November 1974. The IEA was a combination of the most important consumer countries that between them accounted for 80% of the world's oil consumption. It was in fact a consumer front against the producers. In principle, all OECD members could be members of the IEA. Only France declined the privilege, although because of its EC membership Paris nonetheless remained closely involved with the work of the IEA.[90]

Conclusion

With this accord, the adventure of the oil crisis arrived back in port, a journey that had reached an end in every way satisfactory to The Netherlands. With the IEA, an organisation joining forces had come into being, just what The Hague had been advocating for so long. There was no more question of a European power-combine, independent from the USA, being linked to any Euro-Arab dialogue. The Netherlands had played an active role throughout the oil crisis by blocking such an association, as proposed in particular by Paris. Of course, this opposition was not the only reason for the failure of the French objectives.

As we saw earlier, political relations within the EC had by this time undergone remarkable changes. In the three largest member states, power had passed to different political leaders than those at the beginning of the crisis. Compared with their predecessors, the new leaders seemed much keener on preserving good relations with the USA. A period of mounting conflict within the Atlantic world seemed to have passed. The oil crisis, in various respects, had been the nadir of Euro-American tension during this period.

Mutual Western divisions, as was emphasized by Economic Affairs especially, had contributed to a weakening of the Western position vis-à-vis the countries of the Third World and to a growing non-Western self-awareness. The actions of OPEC were in several respects an example and a symbol of this new self-awareness. In the continuation of the OPEC actions, the Group of 77 uncommitted non-Western countries formulated the struggle for a New International Economic Order. It was especially Washington that set itself against this non-Western effort to achieve an international economic redistribution.

Within the Den Uyl government in The Netherlands, there were different opinions on this question. Pronk sympathised with the ambitions of the NIEO. From this perspective he also understood the measures that OPEC had taken. Lubbers, on the contrary, argued for further development of the arrangements made in Washington and was more disposed toward a consumers' front. The Cabinet supported the effort to get the whole energy problem dealt with in the UN General Assembly, but Van der Stoel arrived at this position through more opportunistic, diplomatic considerations than Pronk: this attitude might, after all, contribute to ending the embargo. The strategy worked. Algeria decided in March that the embargo had to be lifted, and as a quid pro quo The Netherlands supported Bouteflika's candidature for the chairmanship of the General Assembly and also attempted to reach agreement with Algeria in the World Bank.

But for the time being there was no lifting of the embargo. It was mainly Saudi Arabia (possibly egged on by Washington) that resisted the decision to lift the embargo. Saudi Arabia and Kuwait declined to vote for abrogation for the specific purpose of thwarting the so-called Euro-Arab dialogue, which had been principally decided to be set up in the context of European Political Cooperation. Effectively blocking this dialogue was a cause for considerable satisfaction on the part of the Americans. It is therefore hardly surprising that The Hague should find American attempts to get the embargo lifted – to say the least – rather ineffective.

During the last phase of the embargo, The Hague played a rather remarkable role, a role which, as was remarked earlier, can be seen as very useful for Dutch interests in the international oil sector. In the Dutch Cabinet, and particularly, in the case of Van der Stoel, there was little feeling for a Euro-Arab dialogue. The latter in fact declined, with not too much hesitation, to adopt the advice given by his civil servants and diplomats to make some gesture toward the Arab countries. The second letter to King Feisal was nonetheless a moment that Van der Stoel's tactics came closest to those of various other EC countries, such as West Germany. But with head held high, Van der Stoel declared in ministerial council on July 12 that The Netherlands 'had not demeaned itself' in the way that certain other European countries most certainly had.[91] It had in fact been neither necessary nor advisable.

The pressure to get the embargo lifted by this stage was slackening. The oil was again arriving in the usual vast quantities in the port of Rotterdam and had already reached its normal level by April. In all probability, even for those countries that had not wanted to lift the embargo, it had only had political significance. In any case, in the spring of 1974 the oil supplies in the Botlek area reached record highs and lay dozens percent higher than in 1973. This is not to say that the decision to lift the embargo on July 10 was not greeted with considerable relief in The Hague. In all, the embargo against The Netherlands had lasted more than eight months.

Conclusion

In many respects, The Netherlands came through the oil crisis rather well. The embargo soon proved ineffective, in the sense that Dutch oil imports at the end of 1973 were no more seriously depleted than those of other Western countries. Indeed, compared with several other West European countries, the oil supply to The Netherlands looked rather healthy. In a book on the oil crisis written from an Arab perspective, the conclusion was correctly drawn that 'friendly states' in Western Europe had suffered more from an embargo laid against The Netherlands than had The Netherlands itself.[1]

Through the months of October 1973 to the spring of 1974, the oil supply to Western Europe as a whole was probably not much lower than during the same period of the previous year. In all West European countries, oil supplies in this period remained more or less at the normal level. In some, partly as a result of the usual increase in winter use, the level dipped: most of all in Belgium, and to a lesser degree in France and Great Britain. The Netherlands in fact came off rather well: stocks in The Netherlands remained virtually normal.[2]

The oil embargo was more a matter of image projection and intimidation. It was mainly the fearful *expectations* that diminished so drastically, while in practice the oil supply itself surprisingly remained at a satisfactory level. This is not to say that there was no cause for alarm. The embargo was undoubtedly an action of a highly threatening kind. Nor were the pessimistic predictions of the Central Planning Bureau lies. Moreover, in December the oil supply to Rotterdam fell by a large percentage over a very short time. It was in these straightened circumstances that the Den Uyl Cabinet adopted drastic measures – the introduction of rationing and the Enabling Act. The oil, however, came rather swiftly back on stream. The Cabinet realised this, which was one of the reasons for the secrecy surrounding the figures relating to the oil supply.

Summarising, we can say that the oil embargo against The Netherlands did not work; and the most important reason for this failure was that the oil companies decided to spread the oil scarcity. In addition, it was from the outset the question of whether those countries participating in the embargo were really prepared to exercise control over the transport of oil to those countries against which the embargo had been imposed. Because the oil companies switched to *rescheduling* the oil supply, it is very well possible that – just as in the USA – oil may have arrived in The Netherlands from embargoing states.

The real sting in the Arab action, however, was the unilaterally declared price increases. These higher prices were to have major consequences for the Dutch economy, though they hit The Netherlands no harder than other oil-importing countries. Moreover, The Netherlands had the additional advantage that the prices of natural gas could, in the short term at least, be linked to the price of oil. Subsequently, therefore, things worked out quite well. Even the fear that the position of Rotterdam as a port and of the Botlek area as a whole might suffer proved to be unfounded. When in July the embargo was eventually lifted, Rotterdam's central position in the international oil sector was unaffected.

As Van der Stoel had maintained in the Council of Ministers, therefore, the Dutch government had not needed to demean itself in the way that some other countries had done. The Cabinet and Van der Stoel, in fact, had with considerable care managed to manoeuvre their way through the crisis. When Voorhoeve singled out Van der Stoel's policy as an example of blatant moral intransigence in the face of Dutch economic interests, this accusation already sounded facile. There had been genuine attempts to conciliate the Arab countries. The second letter to Feisal (see Chapter 8) in fact went so far that it was scarcely distinguishable from the line taken by some other EC member states – such as West Germany – that had been so roundly cursed by The Hague.

The Dutch government, the Foreign Ministry in particular, had to navigate between the demands of the Arab countries on the one hand and those of the Chamber and public opinion on the other. The so-called Thurkow affair should also be seen in this light. But at the same time it has to be said that Van der Stoel certainly did not go as far as some of his advisors and diplomats had wanted. In the main he resisted the pressure openly and explicitly to satisfy the demands of the Arab countries. It was his view, and not without reason, that for The Netherlands to concede to these demands would only lead to further demands.

It was just possible that the Arabs were not primarily concerned with the Dutch policy over the Middle East conflict and Israel at all. Van der Stoel was convinced that the embargo had already been prepared well before October. In his judgement, given the transit function of the Botlek area and Rotterdam, the aim of the embargo against The Netherlands was to put pressure on the whole of Western Europe. The objectives behind the action were political as well as economic.

As we saw earlier, it is not easy to settle on a definitive account of the role and the specific aims of the Arab oil-producing countries. For a start, and this was the view of Dutch diplomats at the time, there were significant differences between the various countries involved; in addition to which the attitude of some of these states changed during the crisis. The oil crisis was therefore a rather complicated power game, which made it particularly difficult for the Den Uyl Cabinet to undertake any action via diplomatic means.

In this regard, it would be interesting to compare the objectives (as well as the self-perception) of the embargo against The Netherlands with that against Denmark. It should also be noted that the Den Uyl government never once attempted to collaborate with this other North-West European victim of the Arab action. At any rate, we have been unable to find anything in the Dutch archives over any such attempt.[3]

But it was not only the Arab countries that were divided among themselves. The same was true of the EC countries. It rapidly became apparent that no-one could count on European solidarity. Within the EC, it was a matter rather of 'every man for himself'; or worse, some partners, France in particular, were possibly exploiting the situation in order to undermine the position of Rotterdam as an oil port. The British government also tried to force the oil companies to leave The Netherlands in the lurch. The oil crisis showed once again that the EC, when it came to the crunch, was not prepared to close ranks.

There were complaints in The Hague over the lack of 'European solidarity', but this was more for public consumption than an expression of real disillusion, for the attitude in the Dutch Council of Ministers was business-like and aimed at the defence of the national economic interest. The government's efforts in this direction were such that oil transit and oil products had to be more sharply reduced than domestic consumption. In case of emergency, the government was even considering requisitioning foreign oil stocks stored in the Botlek area. In any case, the West Germans, Belgians and even the French dependence on Dutch natural gas and Dutch transit were used to persuade these countries to adopt a more pro-Dutch attitude. As a Dutch Foreign Ministry memorandum put it: 'maximum use was made of the uncertainty as to whether the oil destined for transit also fell under the embargo'.[4]

Yet it would be going too far to attribute the relatively satisfactory outcome to the crisis mainly to the performance of the Den Uyl Cabinet. One could equally conclude that the Dutch government and Foreign Minister Van der Stoel clung far too long to the illusion that The Netherlands' EC partners could be brought over to share their view of 'solidarity'; while the pursuit of 'sharing' can only be seen as pointless or even damaging to the Dutch case.

The satisfactory oil supply was in the first place the result of the oil companies' policy, especially of Shell and BP. Once it became apparent that the oil shortage was to be shared out by the oil companies more or less equally over their West European customers, The Hague's need for 'solidarity' rather rapidly evaporated. In December, the Council of Ministers realised that the '*invisible hand*' of the oil companies was rather more useful than the EC partners' '*visible hand*'. Accordingly, there was little disappointment over the less than successful issue of the EC Summit, held in mid-December in Copenhagen.

Some members of the Council of Ministers, not least of whom Den Uyl himself, had difficulties with this dependence on the oil companies.[5] In fact, it was later urged within the Energy Co-ordinating Group that these oil companies should be compelled in whatever way to provide information. And yet the freedom of action of the oil companies proved to be in the Dutch interest. Possible EC interventions, with an eye to the relatively favourable energy situation in The Netherlands and the Dutch production of natural gas, could well have led to disadvantageous complications. In his discussion with Heath, therefore, Den Uyl could not avoid arguing for the maintenance of that freedom, a standpoint that would certainly have astonished any Dutch voter who still remembered the 1972 election manifesto *Turning Point '72*.

The Den Uyl Cabinet therefore, unlike the French, argued for the maintenance of existing arrangements in the international oil sector. This policy fitted the traditional, liberal standpoint that previous Dutch governments had also adopted. Initially, at least, the Council of Ministers argued for some action to be taken by the European Commission, but there was no need for this to be part of any reform-minded EC energy policy. The attempted economic approach to the Arab oil producers, independently of the USA, was also viewed from The Hague with distrust. On the contrary, the Dutch government lent its support to Kissinger's attempts to reassert American leadership by means of an Energy Conference convened in Washington.

The Dutch oil interests were of course the most important reason for this attitude. More generally, there was the fear of growing rivalry and chaos within the world economy. Above all, there was the threat of Western divisions, of the kind seen in the West European 'bilateral oil diplomacy'. Ministers in Den Uyl's Cabinet concerned with the crisis, however, held that American leadership was more important than Western division and impotence. Some even hoped that the United States would make concessions to the Third World, though this turned out to be an illusion, much to the disappointment of Van der Stoel among others.

This did not mean that the American government's position could be blindly trusted in all respects. During the last months of the embargo, the remarkable situation arose that the continuing embargo against The Netherlands could be exploited as a diplomatic means of preventing a closer European approach to the Arab countries. In this situation, the roles in the Arab world began to turn around. The original instigators of the embargo, such as Algeria, now began to urge that it be withdrawn, since continuing the embargo stood in the way of the Euro-Arab dialogue.

There was one clear exception to this general rule of the Dutch Atlantic orientation, and that was the Cabinet's view of the initiative to call a special session of the UN General Assembly. Nixon and Kissinger rejected this idea. Within the Council of Ministers, their scepticism was shared by Lubbers, but the PvdA ministers, Pronk above all, had more sympathy for this endeavour to involve the Third World more. Van der Stoel and the Dutch Foreign Affairs chiefs were of the same mind, albeit out of more opportunistic motives.

Besides, the debates over the special session cannot properly be judged unless we pay far more attention to the major changes that seemed to be taking place in world politics at large. At the beginning of the 1970s, the predominant fear in The Hague – and in the Ministry for Economic Affairs – was that the structure of the post-war world economic order was under pressure. The years of American hegemony, backed by such international organisations as the IMF and the World Bank, seemed to be coming to an end. American–European relations were tense, and the non-aligned nations of the Third World, partly in response to this Western disunity, were demanding a New International Economic Order.

In this respect, the position of the Den Uyl Cabinet was ambiguous. In a certain sense, the Cabinet was itself the product of these changes that seemed to be taking place in the world at this time. Pronk in particular personified the willingness to go some way to meet the demands of the

Third World countries, a readiness which went so far as to generate a certain sympathy for OPEC's actions. Yet most of the Dutch Cabinet members adopted a very cautious stance toward the special session of the General Assembly.

The attitude of the Den Uyl Cabinet toward the oil crisis was certainly not morally inspired, nor was there any question of preaching or of any exalted ideological aims. Any ideological luxury, after all, is scarcely possible when such enormous economic interests are at stake. Neither can we agree with the suggestion of R. Vernon that the Dutch government was 'less knowledgeable, less powerful, and less prepared to act on oil matters' than the British.[6] The manner in which goal-directed power politics were employed, or attempted, by Den Uyl's Dutch Cabinet was just as business-like as that of any other West European government.

Support for Israel might at first sight seem to contradict this assertion. The pro-Israeli attitude of the Den Uyl Cabinet may even have been damaging to Dutch economic interests. During the most ominous days of the oil crisis, the Dutch government was accused of paying too little attention to these interests. Indeed, Den Uyl's government did go rather far in its offer of support to Israel. In secret a considerable quantity of arms and especially munitions, for The Netherlands at least, was delivered: Stemerdink points out that after the deliveries to Israel 'all the depots were empty'. In this action, The Netherlands assumed a strikingly different position from other EC countries. Only from West Germany – in all probability – were American arms and American material transferred to Israel. Moreover, Van der Stoel went further, trying to assist Israel on the diplomatic front, in the first place by blocking a joint French-British EC representation in the Security Council, much to the astonishment of some of his diplomats, such as Fack and Van der Klaauw.

What was the source of this pro-Israeli attitude? One can first of all point to the traditional Dutch bond with Israel. Although this 'special relationship' had become somewhat weaker in the preceding years, these relations undoubtedly played a role in the arms deliveries. In the second place, the land of kibbutzim was popular with the Dutch Labour Party. Van der Stoel's attitude, especially his refusal to accede to the more pro-Arab advice from the top echelon of the Dutch Ministry for Foreign Affairs, can be partly explained by the existing bond between the PvdA and the Israeli Labour Party.[7]

However, in several respects the support for Israel did not represent a departure from the broader objectives of Dutch policy in the face of the oil crisis. We were able to remark that the arms deliveries to Israel took place with the support of the USA and even at the urging of the Americans, even though at the time probably only very few individuals knew this. But more generally, the Dutch Cabinet and Van der Stoel followed the American line on the Middle East. In part, this was a consequence of the usual basic Atlantic assumptions, but this approach meshed perfectly with Dutch interests in the international oil sector, in particular The Hague's rejection of an independent European economic and political rapprochement with the Arab countries over the head of the United States. The Netherlands adopted an Atlantic stance, supporting the Americans' superpower role in the Middle East and rejected a Euro-Arab dialogue as much out of political as out of economic considerations.

In general, the actions of the Den Uyl Cabinet in international affairs can be judged business-like and goal-directed. A similar conclusion, however, would be less justified when it came to domestic measures. Rationing, in particular, was unnecessary. In the first instance, though, government action was effective. During the first phase of the car-free Sundays, the government succeeded in compensating for the reduction in the oil supply by reducing domestic consumption. In this equation, though, it must also be noted that the transit and export of oil and oil products probably declined more sharply than the supply of oil, so that Dutch stocks at the end of the year had not or had hardly been affected.

Reviewing the figures for oil supply, it might justifiably be concluded that there had in fact never been any real supply crisis. The fact that the Cabinet nevertheless took action to reduce the level of domestic consumption is understandable in the light of the threatening attitude the Arab oil exporters appeared to take. Foresight, after all, is the essence of government. Furthermore, the oil supply was in very short order much reduced in December. The measures restricting oil use were therefore not useless. Other measures to deal with the temporary reduction in the oil supply and oil processing – by means of direct talks between those concerned presided over by the Ministry for Economic Affairs – served an important purpose. Later on, the government could happily acknowledge that the setting up of a National Office for Chemical Products had become superfluous.

But it soon became apparent that the embargo was ineffective and that the restriction on the oil supply would be merely temporary. There was oil enough. Before the oil crisis began, the oil companies had shipped extra stocks of oil to The Netherlands, which was no more than normal procedure in connection with the approaching winter. In 1973 there was an additional reason to maximalize stocks: the threatening price increases.

Moreover, rumours had for some time been circulating over a possible 'political use of the oil weapon'. All in all, stocks at the beginning of January 1974 stood at a reassuringly high level,[8] while at the same time signs of recovery in the oil supply were evident. As we have seen, it was also evident within the Dutch Ministry for Economic Affairs that there was no depletion of oil stocks. In fact, shortly afterwards, stocks were found to be higher than the previous year. Rumours were circulating to the effect that oil tankers lay waiting before the coast because storage tanks were overfull.

The introduction of rationing was therefore a superfluous measure, decided at a moment when the oil supply did indeed appear to be waning. The PvdA faction in the Second Chamber later concluded that the government's analysis, set out in the December note 'Restriction of the oil supply and its consequences', was faulty. The conclusion that a 30 to 35% reduction in oil use had to be achieved was a gross exaggeration.[9] By the end of January 1974, there was absolutely no further need to proceed with the whole exercise. No neighbouring country resorted to such measures; the Dutch government here was completely out of step.

Nor were all ministers by that time convinced of the need for rationing. Lubbers, in particular, had his doubts, subsequently dismissed by Rutten, the Economics Affairs secretary-general and chairman of the Co-ordination Group, as 'loss of nerve'.[10] The chief reasons for going through with the plan seem to be all the infrastructural preparation and the concomitant 'credibility' of the Cabinet. In the eyes of top Economics Affairs officials like Rutten, it was mainly the credibility of the Ministry of Economics Affairs that was at stake. Eventually, all Cabinet members agreed to push ahead with rationing, albeit for rather different reasons. Van der Stoel thought calling off rationing would have a provocative effect on the Arab states, since it would amount to a declaration in so many words that the embargo did not work.

During the Cabinet discussions over rationing, it became clear that there was considerable scepticism over the figures furnished by the oil companies. On several occasions, attempts were made to get further supplementary information, for example through the pilot service. There was also mistrust between the different departments involved. Westerterp, apparently, had little faith in the data on oil stocks and supplies used by Economic Affairs (originally supplied by the oil companies).

At such moments, interdepartmental mistrust between the Dutch Ministries of Economic Affairs and Foreign Affairs also surfaced publicly. Before the oil crisis, as we have seen in Chapter 1, Economic Affairs had primacy in matters international arising from the oil problem. During the crisis, however, Foreign Affairs came to play a more prominent role, which was regarded with a somewhat jaundiced eye by the other Ministry. Against this background, it is remarkable that all the reassuring reports on the embargo reaching Foreign Affairs should have played no part in the decision-making over rationing. Willemsen, the chairman of the Co-ordination Group, later expressed his opinion that Rutten considered any doubt cast on the figures furnished by Economic Affairs as an attack on the competence of the Ministry in its management of the consumption-limiting measures.[11]

Rationing lasted but a brief period before ending in chaos and civil disobedience. The oil companies, who had originally urged more far-reaching measures than the car-free Sundays, were partly instrumental in putting the skids under the rationing system. This is not to deny that the measures introduced to reduce consumption were successful up to a point and that they helped to produce the unexpectedly favourable stockpile position. These stocks undoubtedly contributed to the record profits booked by the oil companies in the year 1974. Large stocks at a time of rapidly increasing oil prices could only benefit the companies, a fact which, naturally, did not escape the attention of some members of the Dutch Cabinet. As a former advisor to Den Uyl later put it: 'You don't have to be an economist to see that'.[12]

Although the longer-term consequences of the oil crisis fall beyond the scope of this book, we would nevertheless like to close with a few observations on this aspect. We have already remarked that the oil crisis heralded a turning point in the development of international relations: the failure of the EC's effort to develop a greater political independence from the USA. The oil crisis took place at a remarkable moment. The Western world seemed to find itself in an increasingly weaker position. Above all, the power of the United States appeared to be on the wane. The egregious American setback in Vietnam was an example and a symbol of this. The non-Western countries, both the communist states and the countries of the Third World, had become stronger. The actions of the Arab oil states were a product of these circumstances and seemed only to amplify this weakness in the West.

In the face of these challenges, the West could only respond with discord and division. There were conflicts between the United States and Western Europe. According to the French historian Grosser, 1973 was the andiron of post-war American-European relations. Despite all their best intentions, the EC states were unable to cooperate more effectively (as an alternative to the Pax Atlantica), whether in tackling common energy problems or in relation to the EPC. The oil crisis was merely the proof of this discord.

In the spring of 1974, more pro-Atlantic leaders assumed power in three major EC countries: Wilson, Schmidt and Giscard d'Estaing. This development put an end to the pursuit of an independent EC energy policy. Actions to reach a certain tuning of energy policy within the Western world would subsequently be taken in the Energy Co-ordinating Group and lead to the setting up of the International Energy Agency.

The days of the worst splits in the Western world were past. Recovery of Western unity went hand in hand with the end of European manoeuvring in the Arab direction. There would also be no New International Economic Order. As was remarked at the Dutch Ministry of Economic Affairs, Western discord had been an important cause of the search for a NIEO. The oil crisis was also an important turning point as far as this was concerned. The OPEC measures, both a consequence of and a means to exploit mutual Western rivalries, in the end drove the Western countries back together.

In The Netherlands, the events of 1973 and 1974 were to echo and re-echo for a long time to come. In subsequent years, different lessons were drawn from the whole experience of the oil crisis, one of the more obvious being that in the future, energy matters would have to be treated in a cleaner and more efficient way. Already in the period dealt with here, action was taken to rein in the unrestricted growth of energy consumption. Within a fairly reasonable time, this would lead to the better insulation of homes, among other measures.

In the Hague, the oil crisis ensured a new awareness that the Western oil-consuming countries were vulnerable. In the years after 1973-74, attempts were also made by the Dutch government to improve relations with the Arab world. The Dutch position on the Middle East conflict would slowly but surely shift in a direction more favourable to the Arab countries.[13] In addition, development aid would be offered to Egypt, even though the latter hardly belonged to the poorest of the developing nations. Nonetheless, Egypt was selected in the latter part of 1974 as a land of special help, as a part of a policy of reconciliation by the Dutch government in relation to the Arab world.

Notes

Notes Chapter 1

1 NMFA, Embassy archives Jerusalem (hereafter PAJ), *Code 921.322, Folder MO Conflict I, 8-X-73 thru 31-X 73*, Jongejans to NMFA, no. 26, 09/10/73. See for the plan for a pre-emptive strike: Bar-Siman-Tov, *1987*, pp. 188-194. Meir told Den Uyl later that she was informed about the upcoming attack fifteen hours before this happened. However, she had decided against a pre-emptive strike. See IISG, *Archive Van Thijn, Inv. no. 13*, Diary entry, 12/11/73.

2 See for the battle: Aker, *1985*, pp. 72-105.

3 Garthoff, *1985*, pp. 368-372.

4 This decision was probably also influenced by the wish to turn away the attention from the Watergate scandal. See: Gaddis, *1978* , p. 275.

5 Garthoff, *1985*, pp. 376-382.

6 Lebow and Gross Stein, *1994*, p. 151.

7 Lebow and Gross Stein, *1994*, pp. 165-166.

8 Lebow and Gross Stein, *1994*, pp. 185-186.

9 Grosser, *1978*, p. 350.

10 Soetendorp, *1982*, passim; Schaper, 'Nederland', pp. 229-242 and Grünfeld, *1991*, passim.

11 Hellema, *1990*, passim. In the Netherlands the nationalization of the Suez Canal company was seen as an act which corresponded with the policy of the Indonesian president Sukarno.

12 Schaper, 'Nederland', pp. 235-238.

13 Rusman, 'Dilemma's', pp. 171-180.

14 *Het Parool*, 27/09/93 and Peeters, pp. 208-218.

15 AMD (Archives Ministry of Defence), *Ministeriële Beschikking, no. V96007435*, Memorandum by the Military Intelligence Service (Militaire Inlichtingendienst, MID) about the training of Israeli military at the OCTD, 14/11/95. The three to

four groups consisted of eighteen to twenty persons, with military personnel of various ranks. An Israeli major once flew in from West Germany to attend an exercise. The length of the training, according to MID documents, lasted five weeks at the Doumoulin barracks in Soesterberg.

16 See, *Het Parool*, 25/09/93.

17 Rusman, 'Dilemma's', p. 168.

18 NMFA, *PAJ, Code 921.322, Folder MO-Conflict I, 8-X-73 thru 31- X 73*, Jongejans to NMFA, no. 24, 06/10/73, Van der Stoel to Jerusalem, no. 7. 06/10/73, Jongejans to Den Haag, no. 37589, 07/10/73.

19 NMFA, *Code 999.224, Folder 1250, VN/Midden-Oosten, Part XII*, Van der Stoel to New York, no. 263, 07/10/73.

20 That was the situation before the outbreak of the war.

21 Archives Prime Minister's Office (hereafter *KMP), Beschikking no. 14988*, Memorandum 'Conclusions of the meeting as 8 October', top secret, 09/11/73.

22 NMFA, *Code 999.224, Folder 1250, VN/Midden-Oosten, part XII*, Van der Stoel to New York, no. 266, 08/10/73.

23 NMFA, *PAJ, Code 921.322, Folder "MO-Conflict I, 8-X-73 thru 31- X 73'*, memorandum for the press, 08/10/73.

24 NMFA, *Record Group IV, Code 101, Folder 485, Het Nederlandse beleid tijdens de oliecrisis 1973 - 1977, Volume I* (hereafter: NMFA report on the oil crisis, Part I), p. 66.

25 Grünfeld, *1991*, p. 106. The Directorate-General for Political Affairs (DGPZ) is the political section of the NMFA. DGPZ deals with bilateral relations and security matters.

26 Statement by The Netherlands government regarding the outbreak of the war, 09/10/1973. Italics by the authors.

27 Parliamentary proceedings (*HTK)*, 1973-1974, p. 98.

28 Fack, *1984*, pp. 54-55.

29 NMFA, *Code 913.211.0, Folder Midden-Oosten, Palestijns probleem, Nederlandse standpunten, Volume IV*, Van der Stoel circular, no. 192, 10/10/73.

30 NMFA, *Code 999.224, Folder 1250*, Fack to NMFA, no. 675.841, 09/10/73.

31 NMFA, *PAJ, Code 921.322, Folder MO-Conflict I, 8-X-73 thru 31- X 73*, Jongejans to NMFA, no. 27, 09/10/73.

32 NOVA interview with former Israeli ambassador Chanan Bar On, October 1993.

33 NOVA interview with Bar On, October 1993 and with Van der Stoel, January 1997.

34 Interview with Van den Bergh, February 1997. Within the Labour Party one politician was also brought up to date. It was the parliamentary leader Ed van Thijn. According to him, Vredeling informed him several weeks after the outbreak of the war during a late night visit at his home. Interview with Van Thijn, February 1997.

35 Every American embassy houses a number of officials who work for the *Central Intelligence Agency* (CIA). Such a *CIA station* is run by the Chief of Station (COS) who has his own means of communication with CIA headquarters. For The Netherlands COS: De Graaff and Wiebes, *1998*, passim.

36 Interviews with former CIA officials, Washington, June 1997 and March 1998.

37 Kruimink cannot recall what was discussed in Swift's mansion on that Sunday, but he presumes that it dealt with the ongoing war in the Middle East. Interview with Kruimink, February 1997.

38 According to Kruimink no decision was taken about the arms deliveries in the daily meetings of the Comité Verenigde Inlichtingendiensten Nederland (CVIN) which co-ordinates the Netherlands intelligence activities. Letter of Kruimink to the authors, 11 February 1997.

39 Letter of Van der Stoel to the authors, 26/05/1998 and interview with Vredeling, June 1998.

40 IISG, *Collection Den Uyl, Inv. no. 694*, 'Enkele onderwerpen die van 26 October tot 2 November 1973 in de publiciteit zijn geweest', undated. Another indication can be found in the book by Boon who claimed that two Israeli cargo planes, which landed at Volkel, were forced to return empty. See: Boon, *1976*, p. 283.

41 Peeters, *1997*, p. 198.

42 Interview with Stemerdink, November 1996, and telephone conversation with Stemerdink, June 1998.

43 Interview with Vredeling, June 1998.

44 Peeters, *1997*, p. 198.

45 Interview with Kruimink, March 1997.

46 NMFA, *Code 921.320, Folder 538*, Geheim memorandum van de Secretaris-Generaal van het ministerie van Defensie to de minister van Defensie, 09/10/73.

47 NMFA, *Code 921.320, Folder 538*, Geheim memorandum van de Secretaris-Generaal van het ministerie van Defensie to de minister van Defensie, 09/10/73.

48 Interview with Van der Stoel, January 1997.

49 NMFA, *Code 921.320, Folder 538*, Memorandum DNW to DGPZ, no. 90 ("zeer veel spoed"), 09/10/73.

50 F.E. Kruimink, 'Enkele factoren van belang voor het beoordelen van de vraag, wat is de waarde van 11.000 tankgranaten voor Centuriontanks', undated (but in view of the wording probably 10 October 1973), in IISG, Collection Cees Wiebes.

51 See for the importance of Centurion ammunition: Stemerdink, *1989*, pp. 248-249.

52 The Netherlands refused to criticize openly the French and British stance, even when Israel asked for this. See: NMFA, *BZ-Rapport oliecrisis, Volume I*, p. 78.

53 Kruimink, 'Enkele factoren van belang'.

54 According to Stemerdink and Vredeling the latter had talked on Wednesday morning to Van der Stoel and Den Uyl about the delivery of arms: interviews with Stemerdink and Vredeling, November 1996 and January 1997. Van der Stoel claimed that he only once discussed this issue with Den Uyl and Vredeling. Interview Van der Stoel, January 1997.
55 NMFA, *Code 921.320, Folder 538*, aantekeningen op Memorandum DNW/PC to DGPZ, no. 90 ('zeer veel spoed'), 09/10/73.
56 NMFA, *Code 921.320, Folder 538*, aantekeningen op Memorandum DNW/PC to DGPZ, no. 90, 09/10/73.
57 Interview with Vredeling, January 1997.
58 NOVA interview with Van der Stoel, October 1993. Italics by the authors. A remarkable statement because Meir was on the verge of executing a desperate visit to Washington in order to speed up the US deliveries of arms. Bar-Siman-Tov, *1987*, pp. 206-211.
59 Interview with Van der Stoel, January 1997.
60 Grünfeld, *1991*, p. 65.
61 Grünfeld, *1991*, p. 65.
62 Interview with Stemerdink, November 1996.
63 NOVA interview with Stemerdink, October 1993.
64 Interview with Stemerdink, November 1996.
65 Interview with Stemerdink, November 1996.
66 Telephone conversation with Stemerdink, June 1998.
67 Telephone conversation with Peijnenburg, January 1997.
68 Telephone conversation Meines, January 1997.
69 NOVA interview with Bar On, October 1973.
70 *Het Parool*, 25/09/93.
71 Telephone conversation with Meines, January 1997.
72 Letter of Kruimink to the authors, 11/02/97.
73 Telephone conversation with Stemerdink, June 1998.
74 According to Schulten in *Het Parool*, 25/09/93.
75 NOVA interview with Vredeling, October 1993.
76 Kruimink, 'Enkele factoren van belang'.
77 Kruimink, 'Enkele factoren van belang'.
78 NOVA interview with Bar On, October 1993.
79 NMFA, *Code 921.320, Folder 545 Wapenleveranties aan het Midden-Oosten, 1 July 1973 t/m dec. 1974*, Memorandum DNW to Van der Stoel, via DAM, DGPZ and S, no. 103, 02/11/73.
80 *HTK*, 1973-1974, Aanhangsel, no. 335.
81 Letter of Van der Stoel to the authors, 26/05/98.
82 Telephone conversation with Stemerdink, June 1998.

83 NOVA interview with Van der Stoel, October 1973.
84 NOVA interview with Bar On, October 1973.
85 Interview with Van den Bergh, February 1997.
86 NMFA, *PAJ, Code 921.322, Folder MO-Conflict July thru Oct.*, Letter to Den Uyl, 30/10/73.
87 NOVA interview with Bar On, October 1993.
88 Letter of Lubbers to the authors, 09/02/97.
89 Interview with Kruimink, March 1997.
90 NMFA, *Code 921.320, Folder 538 Wapenleveranties aan Israel, July 1970 to dec. 1974*, anonymous and undated.
91 NMFA, *Code 913.211.0, Folder 5703*, Van Lynden to NMFA, no. 804, 31/10/73.
92 NMFA, *BZ-Rapport oliecrisis, Volume I*, p. 81.
93 NMFA, *Code 921.320, Folder 538*, Van der Stoel to Washington, no. 362, 06/11/73.
94 Interview with Vredeling, January 1997.
95 Interview with Van der Stoel, January 1997.
96 Van der Stoel claims that in those days he presumed that Schlesinger thanked him for the Dutch willingness to open up its airspace for US planes. Letter by Van der Stoel to the authors, 26/05/98.
97 NMFA, *Code 921.320, Folder 545*, Van der Stoel to Paris, 23/10/73.
98 NMFA, *Code 913.211.0, Folder 5704*, Van der Stoel to Gevers, no. 193, 08/11/73.
99 NMFA, *Code 913.211.0, Folder 5704*, Gevers to Van der Stoel, no. 512, 09/11/73.
100 NMFA, *Code 921.320, Folder 545*, Memorandum DAM to S, no. 101, 11/10/73.
101 NOVA Interview with Yamani, October 1973.
102 NMFA, *Code 921.320, Folder 545*, Van Hoeve to NMFA, no. 45, 16/10/73.
103 NMFA, *Code 921.320, Folder 545*, Van der Stoel to Damascus, no. 25, 17/10/73.
104 NMFA, *PA Algiers, Code 614.56, Box 19, Folder Arabische olieboycot*, Van der Stoel to Algiers, 17/10/73.
105 See for this also Chapter 3.
106 NMFA, *BZ-Rapport oliecrisis, Volume I*, p. 69.
107 Grünfeld, pp. 106-107.
108 *KMP, Cabinet meetings*. Minutes of Cabinet Meeting, no. 6181, 12/10/73.
109 NMFA, *BZ-Rapport oliecrisis, Volume I*, p. 71.
110 Interview with Van der Stoel, January 1997.
111 NMFA, *Code 913.211.0, Folder Midden-Oosten, Palestijns probleem, Nederlandse standpunten, Volume IV*, Memo S to M, 17/10/73.
112 NMFA, *PA PVVN New York, Code 999.214 GS, Inv. no. 20, Box 2*, Van der Stoel to PVVN, no. 677, 17/10/73.

113 NFMA, *BZ-Rapport Oliecrisis, Volume I*, p. 81.
114 Grünfeld, *1991*, p. 138.
115 Interview with Van der Klaauw, September 1993.
116 See also Chapter 2.
117 *KMP, Cabinet meetings*. Cabinet Session, no. 6180, 19/10/73.

Notes Chapter 2

1 See for a general overview: Odell, *1986*, pp. 11-50 and Vernon, *1976*, pp. 1-72.
2 Winkler Prins, *1974*, p. 25 and Adelphi Paper no. 117, *1975*, pp. 2-4.
3 Adelphi Paper no. 117, *1975*, p. 3.
4 Aarts, *1996*, pp. 76-78.
5 Bromley, *1991*, pp. 128-129 and Szyliowicz & O'Neill (ed.), *1975*, pp. 52-63.
6 Van Ginkel, *1978*, pp. 25-34.
7 Sutcliffe, *1996*, pp. 202-204.
8 Bromley, *1991*, pp. 136-137.
9 Bromley, *1991*, pp. 138-139.
10 Winkler Prins, *1974*, p. 25 and European Report, *1973*, pp. 3-4.
11 Sherrill, *1983*, pp. 198-199. See also: US National Archives (hereafter NA), Suitland, Maryland, *Nixon Project, WHCF Subject Files, EX TA 1, Box 11*, Memorandum by Love for Nixon, 30/10/73.
12 International Economic Report, *1974*, pp. 46-47 and 106 and Goldstein (ed.), *1982*, p. 73. For Japan: Adelphi Paper no. 115, *1975*, pp. 25-31.
13 Bromley, *1991*, pp. 140-141; Oppenheim, *1976-1977*, pp. 34-43 and International Economic Report, *1974*, p. 48.
14 Bromley, *1991*, pp. 128-129.
15 Yergin, *1991*, pp. 590-591.
16 ASD, *FOIA 8102540*, State Department Memorandum: 'The U.S. and the Impending Energy Crisis', secret, 09/03/72.
17 Yergin, *1991*, pp. 599-602.
18 Schneider, *1983*, p. 38; Van Ginkel, *1978*, pp. 45-50 and Vernon, *1976*, pp. 182-184.
19 Yergin, *1991*, pp. 608-609.
20 What Voorhoeve, *1985*, claims on p. 239.
21 Lieber, *1976*, p. 37.
22 NMFA, *BZ-rapport Oliecrisis, Volume I*, pp. 34-38.
23 NMFA, *BZ-rapport Oliecrisis, Volume I*, p. 41. It concerns the Welsing report, 'De kwetsbaarheid van de Westerse aardolieaanvoer uit het Midden-Oosten en Noord-Afrika in het licht van de Sovjet-politiek', 1971.

24 NMFA, *BZ-rapport Oliecrisis, Volume I*, pp. 41.
25 NMFA, *Code 614.51, Folder 8003*, memorandum PLAN to M, R, T & Z, no. 153, 14/06/73.
26 NMFA, *Code 614.51, Folder 8003*, memorandum PLAN to M, R, T & Z, no. 153, 14/06/73.
27 NMFA, *PA Algiers, Code 614.56, Box 17*, Gevers to NMFA, no. 9282-942/1212, 26/06/73, and Boon to Den Haag, no. 3812/893, 26/06/73.
28 NMFA, *PA Algiers, Code 614.56, Box 17*, Van Lynden to NMFA, no. 680, 20/09/73.
29 AAW, *Code 614.0, Folder Energie: wereldcrisis 1973/74*, Kaufmann to NMFA, no. 524.99.31, 20/07/73.
30 AAW, *Code 614.0, Folder Energie: wereldcrisis 1973/74*, Derksen to NMFA, no. 1919/310, 21/07/73; Hoekman to NMFA, no. 2404/638, 21/08/73, and Renardel to NMFA, no. 1993/528, 06/08/73.
31 Interview with Willemsen, February 1997.
32 *KMP, Cabinet meetings*, minutes of Cabinet meeting, no. 6165, 12/10/73.
33 NMFA, *PA Algiers, Code 614.56, Box 19, Folder Arabisch olie-embargo, 1973*, Van der Stoel to Algiers, no. 209, 24/10/73.
34 NMFA, *Code 613.211.45, Doos 38, Map 440*, Co-ordinator to Van der Stoel, no. 20, top secret, 22/10/73.
35 *HTK*, 1973-1974, pp. 32-322.
36 *KHA*, 09/10/73, p. 714 and Van Ginkel, *1978*, pp. 67-68.
37 NOVA interview with Yamani in Geneva, October 1993. See also the memorandum of a conversation between a Saudi delegation led by Yamani and William P. Rogers in April 1973. ASD, *FOIA 8700809*, Memorandum of Conversation between Yamani, Rogers and other delegates, 16/04/73.
38 Daoudi & Dajani, *1985*, p. 138.
39 Szyliowicz & O'Neill (ed.), *1975*, pp. 63-64 and Adelphi Paper no. 136, *1977*, pp. 2-4. It soon turned out that the FEO had made incorrect appraisals of the situation. Not the embargo but the strict reduction of the oil consumption was the main cause of the problems. The FEO also underestimated the volume of the stocks in the USA and the possibilities to import oil via Canada.
40 AAZ, *Cabinet meetings*, minutes of the Cabinet meeting, no. 6180, 19/10/73. Van der Stoel reported not in this meeting that also Syria had accused The Netherlands of illegal deliveries of arms.
41 NMFA, *PA Algiers, Code 614.56, Box 19, Folder Arabisch olie-embargo, 1973*, Van der Stoel to Algiers, no. 51, 21/10/73.
42 NMFA, *PA Algiers, Code 614.56, Box 19, Folder Arabisch olie-embargo 1973*, Bentinck to NMFA, no. 51, 21/10/73.
43 Coordinator to MP and M, 21/10/73, in: NMFA, *BZ-rapport Oliecrisis, Volume I*, p. 85.

44 NMFA, *Code 613.211.45, Box 38, Folder 440*, Kruimink to Van der Stoel, no. 20, top secret, 22/10/73.
45 NMFA, *Code 913.211.0, Folder 5701*, Van Hoeve to NMFA, no. 49, 22/10/73.
46 See: Adelphi Paper no. 117, *1975*, pp. 26-29.
47 NMFA, *Code 913.211.0, Folder 5701*, Thoe Schwartzenberg to NMFA, no. 3434/847, 29/10/73.
48 *KHA*, 09/11/1973, p. 714.
49 Grünfeld, *1991*, p. 62.
50 NMFA, *Code 614.51, Folder 7944*, Memos by the Shell executive, 26/09/73 and 31/10/73.
51 NMFA, *PA Algiers, Code 614.56, Box 19, Folder Arabisch olie-embargo, 1973*, Van der Stoel circulaire, no. 275, 22/10/73.
52 NMFA, *Code 913.211.0, Folder 5703*, Thoe Schwartzenberg to NMFA, no. 55, 29/10/73 and no. 56, 01/11/73; Scheepers to NMFA, no. 19, 29/10/73 and Schorer to NMFA, no. 20, 30/10/73.
53 NMFA, *Code 913.211.0, Folder 5704*, Gevers to NMFA, no. 499, 02/11/73.
54 NMFA, *PA Algiers, Code 614.56, Box 19, Folder Arabisch olie-embargo, 1973*, Van der Stoel to Djedda, no. 23, 23/10/73.
55 NMFA, *PA Algiers, Code 614.56, Box 19, Folder Arabisch olie-embargo, 1973*, Van der Stoel to Algiers, no. 14, 23/10/73.
56 To this group Shell International also belonged.
57 Grünfeld, *1991*, p. 122. He bases his judgement on the views of Wagner.
58 NMFA, *Code 614.51, Folder 7949*, Memo van der Stoel to S., no. 161/73, 23/10/73.
59 *HEK, Rijksbegroting, Zitting 1973/74, no. 12600, Hoofdstuk V, no. 2*, pp. 13-14, Regeringsverklaring, 23/10/73.
60 NMFA, *Code 913.211.0, Folder 5703*, Memo Van Lynden to M via S, no. 768, 24/10/73.
61 NMFA, *Code 613.211.45, Box 38, Folder 440*, Schorer to NMFA, no. 17, 27/10/73.
62 NMFA, *Code 913.211.0, Folder 5701*, Bentinck to NMFA, no. 55, 24/10/73 and no. 61, 03/11/73.
63 NMFA, *Code 913.211.0, Folder 5701*, Vegelin to NMFA, no. 378, 02/11/73.
64 NMFA, *Code 913.211.0, Folder 5704*, Van der Stoel to Brussels, no. 807/220, 02/11/73.
65 NMFA, *Code 614.51, Folder 7829*, Van Vloten to NMFA, no. 57, 18/10/73.
66 NMFA, *Code 913.211.0, Folder 5703*, Von Oven to NMFA, no. 96, 24/10/73 and no. 104, 01/11/73. Von Oven got the impression that Cairo was satisfied with the cease-fire.
67 NMFA, *Code 913.211.0, Folder 5703*, Van der Stoel to Cairo, no. 47, 30/10/73.

68 NMFA, *Code 913.211.0, Folder 5703*, Van Lunteren to NMFA, no. 57, 24/10/73.
69 In the ultimatum was written: '*from Arab occupied territories*'.
70 NMFA, *Code 913.211.0, Folder 5701*, memo DAM via DGPZ to M, 26/10/73.
71 *KMP, Cabinet meetings*, minutes of Cabinet meeting, no. 6187, 26/10/73.
72 NMFA, *Code 913.211.0, Folder 5703*, Van der Stoel to Teheran, no. 26, 26/10/73, Van der Stoel circulaire, no. 222, 26/10/73 and Van der Stoel to Djedda, no. 26, 27/10/73.
73 *Algemeen Dagblad* and *NRC Handelsblad*, 01/10/73.
74 NMFA, *Code 913.211.0, Folder 5701*, Van der Stoel to Djedda, no. 29 and no. 30, 01/10/73.
75 NMFA, *Code 913.211.0, Folder 5703*, Van der Stoel to Tehran, no. 29, 30/10/73, and Van der Stoel to Tunis, no. 37, 30/10/73.
76 ASD, *FOIA 8903648*, Kissinger to American Embassy, The Hague, no. 215365, 08/11/73.
77 NMFA, *Code 913.211.0, Folder 5704*, Van der Stoel to various embassies, no. 235, 02/11/73.
78 NMFA, *BZ-rapport Oliecrisis, Volume 1*, pp. 96-97.
79 NMFA, *Code 913.211.0, Folder 5703*, Van der Stoel to London, no. 178, 24/10/73.
80 NMFA, *Code 913.211.0, Folder 5703*, Van der Stoel to London, no. 178, 24/10/73.
81 NMFA, *Code 913.211.0, Folder 5703*, Van der Stoel to Djedda, no. 23, 23/10/73 and to Cairo, no. 44, 23/10/73. Also: Van Vloten to NMFA, no. 61, 27/10/73.
82 NMFA, *Code 613.211.45, Box 38, Folder 440*, Renardel to NMFA, no. 60, 25/10/73.
83 NMFA, *Code 613.211.45, Box 38, Folder 440*, Schorer to NMFA, no. 17, 27/10/73.
84 NMFA, *Code 913.211.0, Folder 5703*, Thoe Schwartzenberg to NMFA, no. 55, 29/10/73 and no. 56, 01/11/73; Scheepers to NMFA, no. 19, 29/10/73 and Schorer, Koeweit to NMFA, no. 20, 30/10/73.
85 NMFA, *Code 913.211.0, Folder 5703*, Schorer to NMFA, no. 25, 01/11/73.
86 See for this Chapter 4.
87 Doran made the correct observation that commercial motives were primary for Iraq. See: Doran, *1977*, p. 32.
88 NMFA, *Code 613.211.45, Box 38, Folder 440*, Renardel to NMFA, no. 64, 09/11/73.
89 ASD, *FOIA 8903648*, American Embassy, Tripoli to SecState, no. 1365, 25/10/73.
90 ASD, *FOIA 8903648*, American Embassy Tripoli to Secstate, no. 2396, 30/10/73.
91 NMFA, *Code 912.10 GS, Folder 1007*, Copy memorandum Luns (NATO) to M for DGPZ, Den Haag, 04/11/73.

92 NRC *Handelsblad*, 31/10/73.
93 NMFA, *BZ-rapport Oliecrisis, Volume I*, pp. 109-111.
94 *HTK*, 1973-1974, pp. 524 e.v.
95 NMFA, *Code 913.211.0, Folder 5701,* Van der Stoel circular no. 194, 11/10/73.
96 Interviews with former members of the Mossad and BVD. Also the South-African airline (SAL) is supposed to have flown arms and munitions to Israel via Portugal with cargo planes.
97 NMFA, *Code 921.320, Folder 545*, Van Hoeve to NMFA, no. 45, 16/10/73.
98 NMFA, *Code 613.211.45, Box 38, Folder 440*, Von Oven to NMFA, no. 90, 21/10/73.
99 NMFA, *Code 913.211.0, Folder 5703*, Vroon to NMFA, no. 113, 21/10/73.
100 NMFA, *Code 913.211.0, Folder 5703*, memo VADV, no. 231812/380, 25/10/73.
101 NMFA, *Code 913.211.0, Folder 5703*, Van der Stoel to Damascus, no. 20, 30/10/73.
102 NMFA, *Code 913.211.0, Folder 5702*, Vroon to NMFA, no. 123, 02/11/93 and Van Hoeve to NMFA, no. 59, 03/10/73.
103 NMFA, *BZ-rapport Oliecrisis, Volume I*, pp. 102-103.
104 IISG, *Archive Ed van Thijn, Inv. no. 13, Dairy entry*, 12/11/73, 13/11/73, 15/11/73 and 16/11/73.
105 Terlingen & Roskam, *1997*, pp. 68-73. In *Elseviers Magazine* one could read in November 1973 that Rabbani considered it absolutely unfair to put blame on the Minister. *Elseviers Magazine*, 10/11/73.
106 See for the internal law aspects of the embargo: Moore (ed.), *1977*, pp. 391-445.

Notes Chapter 3

1 Quandt, *1977*, pp. 219-220; Golan, *1976*, pp. 119-121; Sheenan, *1976*, pp. 80-82 and Brecher, *1980*, pp. 302-309.
2 NMFA, *Code 614.51, Folder 8003*, memo Jacobs to chef DES, no. 992/73, 05/11/73, and Daoudi & Dajani, *1985*, p. 139.
3 See for example: Jansen and De Vree, *1985*, pp. 311-317.
4 Gaddis, *1982*, p. 332.
5 Grosser, *1982*, pp. 263-282.
6 Hellema, *1995*, pp. 203-222.
7 Voorhoeve, *1985*, pp. 178-179.
8 See Buchand, *1993*, p. 48, and Urwin, *1995*, pp. 147-149.
9 NMFA, *BZ-rapport Oliecrisis, Volume I*, p. 33.
10 NMFA, *BZ-rapport Oliecrisis, Volume I*, p. 33 passim.
11 NMFA, *Code 996.40, Folder 1827*, nota Van Zutphen over Topconferentie Energiebeleid, 16/10/73.

12 NMFA, *BZ-rapport Oliecrisis, Volume I*, pp. 43-44.

13 AEZ, CAB, *Afdeling PAZ, Sectie DG Energie, Box 85-0947, Folder 2*, Werkgroep voor Economisch, Sociologisch and Statistisch Onderzoek van het Rijnmond-gebied (ESSOR), nota inzake het te voeren beleid n.a.v. de oliecrisis, 14/02/74, p. 19.

14 NMFA, *Code 913.211.0, Folder 5703*, De Ranitz to NMFA, no. 619 and 622, 31/10/73.

15 Costigliola, *1992*, pp. 178-179.

16 Agulhon, *1993*, p. 439 and Hanrieder & Auton, *1980*, p. 151.

17 Grosser, *1978*, p. 354.

18 Simonian, *1985*, 1985, p. 208, and Roussel, *1984*, p. 475.

19 NMFA, *Code 996.40, Folder 1827*, memo DIE to M and T via S, no. 927/73, 01/11/73.

20 NMFA, *Code 614.51, Folder 7955*, Boon to NMFA, no. 962.198, 04/10/73.

21 NMFA, *Code 913.211.0, Folder 5703*, Gevers to NMFA, no. 494, 01/11/73, and Turner, *1980*, p. 178.

22 Yergin, *1991*, pp. 623-624.

23 NMFA, *BZ-rapport Oliecrisis, Volume I*, p. 120.

24 NMFA, *Code 913.211.0, Folder 5703*, Gevers to NMFA, no. 498, 02/11/73.

25 NMFA, *PA Algiers, Box 19, Folder 614.56*, Van der Stoel to Algiers, 03/11/73.

26 NMFA, *Code 913.211.0, Folder 5704*, Gevers to NMFA, no. 505, 06/11/73, and Van der Stoel to London, no. 193, 07/11/73.

27 NMFA, *BZ-rapport Oliecrisis, Volume I*, p. 108.

28 Bark & Gress, *1993*, pp. 260 and 281 and NMFA, *PA Bonn, Code 614.51, Box 201*, Van der Stoel to Bonn, no. 189, 01/11/73.

29 NMFA, *Code 614.51, Folder 7835*, memo DIE to M and T and R, 02/11/73.

30 NMFA, *Code 913.211.0, Folder 5704*, Kupers to NMFA, no. 420, 02/11/73.

31 NMFA, *Code 913.211.0, Folder 5703*, De Roos to NMFA, no. 74, 30/10/73.

32 NMFA, *Code 913.211.0, Folder 5704*, Jongejans to NMFA, no. 3901/806, 01/11/73.

33 NMFA, *Code 614.51, Folder 7859*, Krijgsman to NMFA, no. 5175, 25/10/73.

34 NMFA, *Code 913.211.0, Folder 5703*, Van Schelle to NMFA, no. 371 & 372, 29/10/73.

35 NMFA, *Code 614.51, Folder 7859*, Krijgsman to NMFA, no. 9345, 30/10/73.

36 NMFA, *Code 996.236.3, Folder 201*, memo DES to M via DGES and S, no. 995, 26/10/73.

37 Van Lennep, *1991*, p. 247.

38 NMFA, *Code 996.236.3, Folder 201*, memo DES to M via DGES and S, no. 995, 26/10/73.

39 NMFA, *BZ-rapport Oliecrisis, Volume I*, pp. 117-119.

40 NMFA, *Code 613.211.45, Folder 440, Midden-Oosten, olieboycot, Volume I*,

Van der Stoel to Washington, 31/10/73 and 04/11/73. When Van der Stoel learned about this, he immediately contacted his ambassador and urged him to approach Kissinger. Van der Stoel would appreciate it highly if the American was willing to promote the Dutch interests during his upcoming visit to the Middle East.

41 NMFA, *Code 996.40, Folder 1847*, memo DIE to DGES, no. 977/73, 01/11/73.

42 *KMP, Cabinet meetings*, minutes of Cabinet meeting, no. 6201, 01/11/73. The ministers concluded that apart from the EC there were other possibilities left to use. According to Vredeling the NATO stocks in The Netherlands could be used. Van der Stoel confirmed that the oil pipeline system of NATO was also very important.

43 NMFA, *BZ-rapport Oliecrisis, Volume I*, pp. 121-122.

44 NMFA, *Code 996.40, Folder 1827*, memo DIE to M and T via S, no. 927/73, 01/11/73.

45 NMFA, *BZ-rapport Oliecrisis, Volume I*, p. 105.

46 Voorhoeve, *1985*, pp. 237-238 and NMFA, *BZ-rapport Oliecrisis, Volume I*, p. 130.

47 NMFA, *BZ-rapport Oliecrisis, Volume I*, p. 132.

48 NMFA, *BZ-rapport Oliecrisis, Volume I*, p. 133.

49 NMFA, *PA Parijs, GS, Code 912.1, Box 2*, Van der Stoel circular, no. 243, 08/11/73, and IISG, *Collection Den Uyl, inv. no. 1111*, Memorandum regarding the deliberations on the oil crisis in the EC, undated.

50 Grünfeld, *1991*, p. 76.

51 IISG, *Collection Ed van Thijn, inv. no. 13, Diary entry*, 08/11/73.

52 NMFA, *PA Parijs, GS, Code 912.1, Box 2*, Van der Stoel circular, no. 243, 08/11/73.

53 Letter by Van der Stoel to the authors, 25/03/98.

54 Lieber, *1976*, p. 14.

55 NMFA, *PA Parijs, GS, Code 912.1, Box 2*, Van der Stoel circular, no. 243, 08/11/73.

56 Yergin, *1991*, p. 628.

57 Kissinger, *1982*, p. 713, and Stuart & Tow, *1990*, pp. 81-82.

58 Simonian, *1985*, p. 205.

59 Kissinger, *1982*, p. 718, and *de Volkskrant*, 06/11/73.

60 *Het Parool*, 06/11/73.

61 *Trouw*, 06/11/73.

62 *NRC Handelsblad*, 06/11/73.

63 *NRC Handelsblad*, 09/11/73.

64 *KMP, Cabinet meetings*, minutes of Cabinet meeting, no. 6206, 09/11/73.

65 *KMP, Cabinet meetings*, minutes of Cabinet meeting, no. 6206, 09/11/73.

66 'The use of the oil weapon speeded up and crystallized the EEC position, rather than fundamentally changing it'. In: Adelphi Paper no. 117, *1975*, p. 8.
67 Interview with Rutten, January 1997.

Notes Chapter 4

1 NRC *Handelsblad*, 20/10/73.
2 *Het Parool*, 24/10/73.
3 Archives Ministry for General Affairs (hereafter AAZ), *RVD-Dossier Coördinatiegroep Voorlichting Energietekort*, Van de Graaf to Den Uyl, no. 595, 29/10/73.
4 Lieber, *1976*, p. 9.
5 Archives Ministry of Economic Affairs (hereafter AEZ), *CAB, Dept. PAZ, Section DG Energy, Box 85-0947, Folder 2, Werkgroep voor Economisch, Sociologisch en Statistisch Onderzoek van het Rijnmond-gebied (ESSOR)*, memorandum regarding policy in view of oil crisis, 14/02/74, p. 7.
6 Mitchell, *1989*, p. 430.
7 Mitchell, *1989*, pp. 482-484 and Winkler Prins, *1974*, p. 122.
8 AEZ, *CAB, Dept. PAZ, Section DG Energy, Box 85-0947, Folder* 2, Werkgroep voor Economisch, Sociologisch and Statistisch Onderzoek van het Rijnmond-gebied (ESSOR), memorandum regarding policy in view of oil crisis, 14/02/74, p. 7.
9 NRC *Handelsblad*, 26/10/73.
10 *De Volkskrant*, 27/10/73.
11 AEZ, *RBA, Box 20-0006, Folder no.* 3, Directoraat-Generaal voor Energievoorziening, report by J.C.W. de Vries, 09/11/73.
12 AEZ, *Dept. PAZ, Box 251, Folder 30*, Van Eupen to all oil companies, no. 373/8856/EA, 16/10/73.
13 Grünfeld, *1991*, p. 100.
14 Interview with Willemsen, February 1997.
15 *KMP, Cabinet meetings*, minutes of Cabinet meeting, no. 6180, 19/10/73.
16 AEZ, *Directorate AEP, Archives College van Directeuren-Generaal, Box 99, Folder Stafbesprekingen met de minister AEP*, conclusions of conversation with Lubbers et al., 22/10/73.
17 NMFA, *Code 614.51, Folder 8003*, Memo DES to Van der Stoel, no. 957, 30/10/73.
18 For example from Lybia: NMFA, *Code 913.211.0, Folder 5703*, Van der Stoel to Washington, no. 354, 31/10/73.
19 NMFA, *Code 913.211.0, Folder 5703*, De Ranitz to Den Haag, no. 619 and no. 622, 31/10/73.

20 Rosenthal and Scholten, *1977*, p. 112.

21 Interview with Rutten, January 1997, and with Willemsen, February 1997.

22 *Trouw*, 20/10/73.

23 NRC *Handelsblad*, 24/10/73.

24 *De Volkskrant*, 31/10/73.

25 AEZ, *RBA, Box 20-0006, Folder no. 3*, Directoraat-Generaal voor de Energievoorziening, report by J.C.W. de Vries, 09/11/73.

26 AEZ, *Directorate AEP, Archives College van Directeuren-Generaal, Box 99, Folder Stafbesprekingen met de minister AEP*, Conclusions of conversation with Lubbers et al., 29/10/73.

27 AEZ, *RBA, Box 20-0009, Folder 2*, Rapport inzake de aardoliedistributie in de periode 19 oktober 1973 tot 28 februari 1974, pp. 3a-12.

28 *HTK*, 1973-1974, pp. 528-530.

29 NMFA, *BZ-rapport Oliecrisis, Volume I*, p. 115.

30 *HTK*, 1973-1974, pp. 527-528.

31 *HTK*, 1973-1974, pp. 528-540.

32 NMFA, *Code 614.51, Folder 8003*, Eerste rapport Coördinatiegroep Maatregelen Oliecrisis, 08/11/73.

33 KMP, *Cabinet meetings*, minutes of Cabinet meeting, no. 6159, 02/11/73.

34 KMP, *Cabinet meetings*, minutes of Cabinet meeting, no. 6195, 02/11/73.

35 AEZ, *Archive bureau Secretaris-Generaal, Box 294, Folder 2, Oliecrisis*, Van Agt to Lubbers, 07/11/73.

36 *De Telegraaf*, 05/11/73.

37 *De Tijd*, 05/11//73.

38 *Haagsche Courant*, 05/11/73.

39 *Het Parool*, 31/10/73.

40 NRC *Handelsblad*, 03/11/73.

41 NMFA, *Code 614.51, Folder 8003*, Eerste rapport Coördinatiegroep Maatregelen Oliecrisis, 08/11/73 and letter from W.Q.J. Willemsen to authors, 26 May 1998.

42 AAZ, *RVD-Dossier Coördinatiegroep Voorlichting Energietekort*, memo Coördinatiegroep Maatregelen Oliecrisis, 08/11/73.

43 AMJ (Archives Ministry of Justice), *File A 73/099, Folder energiecrisis, Part I*, Uittreksel uit PG-vergadering, 08/11/73.

44 AMJ, *File A 73/099, Folder energiecrisis, Part II*, Nota voor de Ministerraad over de aardoliecrisis, 08/11/73.

45 AMJ, *File A 73/099, Folder Energiecrisis, Part I*, Uittreksel PG-vergadering, 08/11/1973.

46 AEZ, *Directorate AEP, Archive College van Directeuren-Generaal, Box 99, Folder Stafbesprekingen met de minister*, conclusions of conversation with Lubbers et al., 05/11/73.

47 Grünfeld, *1991*, p. 93. Grünfeld refers to a meeting of the politburo of the Labour Party on 12 November 1973.
48 Rosenthal and Scholten, *1977*, p. 103.
49 NMFA, *PA Bonn, Code 614.51, Box 201*, Van der Stoel to Bonn, no. 189, 01/11/73.
50 Interview with Willemsen, February 1997.
51 Grünfeld, *1991*, p. 76.
52 Yergin, *1991*, pp. 619-620, and Van Seumeren, *1989*, p. 110.
53 Tweede-Kamerfractie Partij van de Arbeid, 'Onderzoek naar de rol van de oliemaatschappijen in de Oliecrisis van 1973-1974', June 1975, pp. 10-11.
54 Interview with Wagner, March 1997, and NMFA, *Code 996.236.3, OESO Oliecommissie, Folder 201, Part III*, Memorandum of DES to M via DGES, 06/11/73.
55 NMFA, *Code 996.236.3, OESO Oliecommissie, Folder 201, Part III*, Memorandum of DES to M via DGES, 06/11/73.
56 NMFA, *Code 996.236.3, OESO Oliecommissie, Folder 201, Part III*, Memorandum of DES to M via DGES, 06/11/73. See: Van Seumeren, *1989*, p. 113.
57 Sampson, *1975*, p. 263.
58 NMFA, *Code 996.236.3, Folder 201*, memo DES to DGPZ, no. 996, 07/11/73.
59 Interview with Wagner, March 1997.
60 Adelphi Paper no. 117, *1975*, p. 7; Heikal, *1975*, pp. 273-274 and Church, *1977*, pp. 39-40.
61 NMFA, *Code 614.51, Folder 8003*, memo DES to Van der Stoel, no. 957, 30/10/73.
62 NMFA, *Code 614.51, Nederlandse maatregelen n.a.v. de oliecrisis 1973/74, Folder 8003*, Memorandum DES to Van der Stoel, no. 1006, 09/11/73. The whole issue of transfer of oil will be dealt with in Chapter 6.
63 Interview with Willemsen, February 1997.
64 Letter by W.Q.J. Willemsen to authors, 26 May 1998.
65 NMFA, *Code 614.51, Folder 8003*, Eerste rapport Coördinatiegroep Maatregelen Oliecrisis, 08/11/73.
66 AEZ, *RBA, Box 20-0003, File no. 7*, Letters E.F. Geessink, ministerie van Landbouw and Visserij to EZ, no. ACB 612, 06/11/73 and no. ACB 637, 20/11/73.

Notes Chapter 5

1 NMFA, *Code 913.211.0, Folder 5704*, De Roos to NMFA, no. 80, 08/11/73.
2 NMFA, *Code 613.211.45, Box 38, Folder 440*, Von Oven to NMFA, no. 118, 17/11/73.

3 NMFA, *Code 913.211.0, Folder 5705,* Von Oven to NMFA, no. 122, 21/11/73.
4 NMFA, *Code 613.211.45, Box 38, Folder 440*, Van der Stoel to Cairo, no. 56, 19/11/73.
5 NMFA, *BZ-rapport Oliecrisis, Volume II,* pp. 149-152.
6 NMFA, *Code 913.211.0, Folder 5704,* Schorer to NMFA, no. 35, 12/11/73.
7 NMFA, *Code 614.51, Folder 7917,* De Ranitz to NMFA, no. 668, 15/11/73.
8 NMFA, *Code 913.211.0, Folder 5705,* memo DEU to DAM, No.167, 19/11/73.
9 NMFA, *PA Bonn, Code 614.51, Box 5,* memo De Beus, no. 401, 06/11/73.
10 NMFA, *Code 614.51, Folder 8004,* Von Oven to NMFA, no. 106, 02/11/74.
11 NMFA, *Code 614.51, Folder 8008,* Boon, Rome to NMFA, no. 276, 27/11/73.
12 NMFA, *Code 614.51, Folder 8004,* Boon, Rome to NMFA, no. 302, 13/12/73, and Van Lynden, Washington to NMFA, no. 936, 28/12/73.
13 NMFA, *Code 614.51, Folder 8004,* Baghdad to NMFA, no. 4013/993, 29/12/73.
14 Grünfeld, *1991*, p. 80.
15 NMFA, *Code 913.211.0, Folder 5705,* Van der Stoel to various embassies, no. 255, 18/11/73. See also: Netherlands National Archives, hereafter: NNA), *Collection Van Roijen, 2.21.183, Inv. no. 62*, Van der Stoel to Van Roijen, 30/11/73, and memo about conversation with Yamani, undated.
16 NMFA, *Code 613.211.45, GS, Folder 440, Midden-Oosten: olieboycot, Volume I*, Boon to NMFA, no. 253, 14/11/73.
17 Interview with Van der Stoel, January 1997.
18 NMFA, *Code 912.1 GS, Folder 1026*, Derksen to NMFA, no. 67, 18/11/73.
19 NNA (Dutch National Archive), *Collection Van Roijen, 2.21.183, Inv. no. 62*, Queen Juliana to King Feisal plus memorandum, 06/02/74. The reason for this late reply is unknown to the authors.
20 NMFA, *Code 913.211.0, Folder 5704,* memo DAM to M via DGPZ, no. 93, 22/11/73.
21 NMFA, *Code 913.211.0, Folder 5705,* De Vreede to NMFA, no. 33, 23/11/73.
22 NMFA, *Code 913.211.0, Folder 5704,* Van der Stoel to various embassies, no. 29 and 59, 23/11/73.
23 *HTK*, 1973-1974, 12.600, V, no. 14.
24 *HEK*, 1973-1974, pp. 111-112.
25 NMFA, *BZ-rapport Oliecrisis, Volume II*, p. 154.
26 NMFA, *BZ-rapport Oliecrisis, Volume II*, p. 154.
27 Letter of Van der Stoel to the authors, 25 March 1998.
28 NMFA, *BZ-rapport Oliecrisis, Volume II*, p. 160.
29 *Het Parool* and *de Volkskrant*, 06/12/73.
30 NNA, *Archive De Koster, Inv. no. 2.21.291, Folder 80*, note by H.J. de Koster, 18/12/73.
31 *HTK*, 1973-74, p. 1000.

32 NMFA, *Code 913.211.0, Folder 5704*, De Ranitz to NMFA, no. 655, 08/11/73.
33 IISG, *Collection Ed van Thijn, Inv. no. 13, Diary entry*, 07/11/73.
34 NMFA, *Code 913.211.0, Folder 5701*, De Ranitz to NMFA, no. 703, 27/11/73.
35 NMFA, *Code 913.211.0, Folder 5701*, Gevers to NMFA, no. 539, 23/11/73.
36 NMFA, *Code 913.211.0, Folder 5701*, Gevers to NMFA, no. 516, 12/11/73; IISG, *Collection Den Uyl, Inv. no. 1138*, Memorandum of conversation Den Uyl-Heath, 11/11/73, and *Collection Ed van Thijn, Inv. no. 13, diary entry*, 08/11/73.
37 *KMP, Cabinet meetings*, minutes of Cabinet meeting, Council for Economic Affairs, no. 6115, 16/11/73 and IISG, *Collection Den Uyl, Inv. no. 1138*, Memorandum of conversation Den Uyl-Heath, 11/11/73. See for the British problems also: Wilson, *1979*, pp. 22-33.
38 NMFA, *Code 913.211.0, Folder 5701*, De Beus to NMFA, no. 443, 16/11/73.
39 NMFA, *Code 996.40, Folder 1847*, Sassen to NMFA, no. 200, 19/11/73.
40 NMFA, *Code 996.236.3, Folder 201*, Kaufmann to NMFA, no. 95, 21/11/73.
41 NMFA, *BZ-rapport Oliecrisis, Volume II*, p. 174.
42 *KMP, Cabinet meetings*, minutes of Cabinet meeting, no. 6226, 23/11/73. See for Kissinger's irritation: Kissinger, *1982*, p. 719.
43 NMFA, *BZ-rapport Oliecrisis, Volume II*, pp. 145-146.
44 NMFA, *PA Algiers, Code 614.56, Box 17*, Van Hoeve to NMFA, no. 73, 22/11/73, and Bentinck to NMFA, no. 78, 24/11/73.
45 NMFA, *Code 613.211.45, Box 38, Folder 440*, De Ranitz to NMFA, no. 704, 27/11/73.
46 NMFA, *Code 913.211.0, Folder 5701*, Bentinck to NMFA, no. 73, 17/11/73.
47 NMFA, *Code 913.211.0, Folder 5704*, Van der Stoel circular, no. 265, 22/11/73.
48 Interview with Van der Stoel, January 1997, and letter by Lubbers to the authors, 9 February 1997.
49 NMFA, *Code 614.51, Folder 8003*, memo DES, no. 1007, 09/11/73.
50 NMFA, *Code 913.211.0, Folder 5704*, Van der Stoel to various embassies, no. 42, 27/11/73.
51 NMFA, *Code 613.211.45, Box 38, Folder 440*, Van Schelle to NMFA, no. 408, 27/11/73.
52 *KMP, Cabinet meetings*, minutes of Cabinet meeting, no. 6234, 30/11/73.
53 Van Elslande declared that Israel should withdraw from *all* occupied territories. Later Belgium was to be rewarded for this statement when it was 'transferred' to the category of 'friendly states'.
54 NMFA, *Code 913.211.0, Folder 5705*, Memo Vegelin van Claerbergen to DGPZ, 30/11/73.
55 Letter of Lubbers to the authors, 9 February 1997.
56 NMFA, *Code 913.211.0, Folder 5701*, Van der Stoel circular, no. 277, 03/12/73. Lubbers denied that he declared that The Netherlands was an opponent of annexation. Letter Lubbers to the authors, 9 February 1997.

57 *Trouw*, 04/12/73.

58 NMFA, *Code 613.211.45, Box 38, Folder 440*, Van der Stoel to Washington, no. 380, 22/11/73 and IISG, *Collection Ed van Thijn, Inv. no. 13, Diary entry*, 08/11/73.

59 NMFA, *BZ-rapport Oliecrisis, Volume II*, p. 167.

60 NMFA, *Code 613.211.45, Box 38, Folder 440*, Van der Stoel to Washington, no. 380, 22/11/73.

61 IISG, *Collection Den Uyl, Inv. no. 1110*, Transcript of a bilateral conversation between Foreign Affairs and Economic Affairs about the oilcrisis, 26/11/73.

62 *KMP, Cabinet meetings*, minutes of Cabinet meeting, no. 6226, 23/11/73.

63 NMFA, *GS, Code 912.1, Folder 1041*, Van Lynden to NMFA, no. 870, 30/11/73.

64 NMFA, *Code 913.211.0, Folder 5700*, memo PLAN to Van der Stoel, no. 210, 05/12/73.

65 NMFA, *Code 913.211.0, Folder 5700*, Van Lynden to Van der Stoel, no. 877, 30/11/73.

66 NMFA, *Code 913.211.0, Folder 5700*, memo DWH to DGPZ, 07/12/73.

67 NMFA, *Code 913.211.0, Folder 5703*, Scheltema to S, 05/12/73.

68 NMFA, *Code 913.211.0, Folder 5704*, De Ranitz to NMFA, no. 741, 08/12/73.

69 Interview with Helfrich, February 1997.

70 NMFA, *Code 912.2, Folder 3581*, De Ranitz to NMFA, no. 727, 03/12/73.

71 NMFA, *Code 614.51, Folder 7917*, De Ranitz to NMFA, no. 738, 07/12/73.

72 Marshall, *1997*, p. 90.

73 NMFA, *Code 913.211.0, Folder 5701*, De Beus to NMFA, 19/11/73.

74 Simonian, *1985*, pp. 206-210.

75 *KMP, Cabinet meetings*, minutes of the Cabinet meeting, *no.* 6226, 23/11/73.

76 NMFA, *Code 913.211.0, Folder 5704*, Van Schelle to NMFA, no. 854.937, 01/12/73.

77 NMFA, *Code 913.211.0, Folder 5704*, Boon to NMFA, no. 962.398, 04/12/73 and no. 962.400, 05/12/73.

78 NMFA, *Code 614.51, Folder 7955*, Boon to NMFA, no. 297, 13/12/73.

79 NMFA, *BZ-rapport Oliecrisis, Volume I*, p. 120.

80 NNA, *2.02.05.02, Archives of the Cabinet Office, Inv. no. 1225*, Note Ringnalda to Den Uyl, no. 547, 02/11/73.

81 NMFA, *Code 913.211.0, Folder 5704*, Van der Stoel to various embassies, no. 208, 27/11/73.

82 Marshall, *1997*, p. 90, and *KMP, Cabinet meetings*, minutes of Cabinet meeting, no. 6115, 16/11/73.

83 *KMP, Cabinet meetings*, minutes of Cabinet meeting, *no.* 6226, 23/11/73.

84 IISG, *Collection Den Uyl, Inv. no. 1110*, Transcript of a bilateral meeting between Foreign Affairs and Economic Affairs about oil, 26/11/73.

85 NMFA, *BZ-rapport Oliecrisis, Volume II*, pp. 170-172.
86 NMFA, *BZ-rapport Oliecrisis, Volume II*, pp. 170-172.
87 Grünfeld, *1991*, p. 83.
88 Grünfeld, *1991*, p. 83, incorrectly writes that The Netherlands had high hopes about Copenhagen.
89 *KMP, Cabinet meetings*, minutes of Cabinet meeting, Council for European Affairs, 05/12/73.
90 NMFA, *BZ-rapport Oliecrisis, Volume II*, pp. 176-177.
91 NMFA, *BZ-rapport Oliecrisis, Volume II*, p. 180.
92 *KMP, Cabinet meetings*, minutes of Cabinet meeting, Council for European Affairs, 05/12/73.
93 AEZ, *Archive DG-BEB, no. 89-1227, Box 3283, Folder 7*, Memorandum by the Minister of Economic Affairs, 05/12/73.
94 *KMP, Cabinet meetings*, minutes of Cabinet meeting, Council for Economic Affairs, no. 6279, 06/12/73. See for all memoranda: IISG, *Collection Den Uyl, Inv. no. 1110*.
95 *KMP, Cabinet meetings*, minutes of Cabinet meeting, Council for Economic Affairs, no. 6279, 06/12/73.
96 NMFA, *BZ-dossier Oliecrisis, Volume II*, p. 178.
97 NMFA, *BZ-dossier Oliecrisis, Volume II*, pp. 182-183.
98 NMFA, *BZ-dossier Oliecrisis, Volume II*, pp. 161-163 and interview with Van der Stoel, January 1997.
99 NMFA, *BZ-dossier Oliecrisis, Volume II*, pp. 161-163; Campbell, *1993*, p. 558 and Simonian, *1985*, pp. 211-212.
100 Hellema, *1995*, pp. 211-217.
101 NMFA, *BZ-dossier Oliecrisis, Volume II*, pp. 178-179 and IISG, *Collection Den Uyl, Inv. no. 1139*, Memorandum Main Issues Summit, undated and letter Ortoli, 12/12/73.
102 *De Volkskrant*, 14/12/73.
103 NMFA, *Code 913.211.0, Folder 5704*, Van der Stoel circular, no. 288, 17/12/73.
104 Brandt, according to Van der Stoel, was much firmer compared with Scheel. Until then Scheel had switched sides constantly between Bonn and Paris. Interview with Van der Stoel, January 1997.
105 Simonian, *1985*, pp. 213-214.
106 Jørgensen, *1989*, pp. 208-210.
107 'Boebie, die oliecrisis gaat toch niet kwakkelen, hè', in: *HP/De Tijd*, 02/05/97, and Jørgensen, *1989*, p. 210.
108 NMFA, *BZ-dossier Oliecrisis, Volume II*, pp. 184-188 and Campbell, *1993*, p. 559.
109 IISG, *Collection Den Uyl, Inv. no. 1139*, communiqué European Summit, undated.

110 Simonian, *1985*, pp. 200 and 215-216. For other monetary and EC matters which were discussed in Copenhagen: pp. 217-224. See also: Joustra & Van Venetië, *1989*, p. 131.
111 NMFA, *Code 912.1, Folder 2315*, Memo Van der Stoel to S, 20/12/73. London profited enormously from this regional support. See. Gann & Duignan, *1998*, pp. 145-146.
112 *Het Parool*, editorial comment, 22/12/73.
113 *Trouw*, editorial comment, 20/12/73.
114 *NRC Handelsblad*, editorial comment, 16/12/73.
115 IISG, *Collection Den Uyl, Inv. no. 1139*, Note by H. van den Bergh, no. F/434, February 1974.

Notes Chapter 6

1 AEZ, *Archive RBA, Box 20-0009, Folder 5*, Tweede rapport Coördinatiegroep Maatregelen Oliecrisis, 14/11/73.
2 AEZ, *Archive RBA, Box 20-0009, Folder 5*, Tweede rapport Coördinatiegroep Maatregelen Oliecrisis, 14/11/73.
3 *KMP, Cabinet meetings*, minutes of Cabinet meeting, no. 6206, 09/11/73.
4 *KMP, Cabinet meetings*, minutes of Cabinet meeting, no. 6115, 16/11/73 and no. 6226, 23/11/73.
5 AEZ, *Archive PAZ, Box 616, Folder 2*, Verslag 4e vergadering EZ Werkgroep Oliecrisis, 21/11/73.
6 Interview with Helfrich, February 1997. When in late November a meeting took place with some members of the Cabinet, the oil companies, according to Helfrich, opposed distribution. That would be impossible solely in The Netherlands. However, Den Uyl stuck to his original stance. 'And on such occasion one can only salute and comply', according to Helfrich.
7 AEZ, *Archive RBA, Box 20-0009, Folder 5*, Derde Rapport Coördinatiegroep Maatregelen Oliecrisis, 22/11/73.
8 NOVA interview with Westerterp, October 1993.
9 *KMP, Cabinet meetings*, minutes of Cabinet meeting, no. 6226, 23/11/73.
10 AEZ, *Archive PAZ, Box 616, Folder 2*, Verslag 5de vergadering EZ Werkgroep Oliecrisis, 28/11/73.
11 NMFA, *Code 613.211.45, Box 38, Folder 440*, memo Rutten to Van der Stoel, 22/11/73.
12 *KMP, Cabinet meetings*, minutes of Cabinet meeting, no. 6234, 30/11/73.
13 NOVA interview with Westerterp, October 1993.
14 *KMP, Cabinet meetings*, minutes of Cabinet meeting, Council for Economic Affairs, 05/12/73.

15 KMP, *Cabinet meetings*, minutes of Cabinet meeting, no. 6267, 10/12/73.

16 AAZ, RVD-*Dossier Coördinatiegroep Voorlichting Energietekort*, Rutten to Lubbers, plus the CEC note 'Beperking van de aanvoer van ruwe aardolie naar Nederland: globale initiële gevolgen en perspectieven voor 1974', 04/12/73.

17 KMP, *Cabinet meetings*, minutes of the Cabinet meeting, no. 6206, 09/11/73. Economic Affairs had made that decision with the exception of the Benelux. Brussels had in the meantime also licensed oil exports inside the Benelux. This was in contradiction with the Benelux Treaty. Economic Affairs intended to retaliate. One should not forget that the exports to Belgium were bigger compared to the imports.

18 AEZ, *Archive* RBA, *Box 20-0006, Folder 3*, Memorandum by De Vries (DGE), 09/11/73.

19 NMFA, *Code 614.51, Nederlandse maatregelen t.a.v. de oliecrisis 1973/74, Folder 8003*, memo DES to Van der Stoel, no. 1006, 09/11/73.

20 NMFA, *Code 614.51, Nederlandse maatregelen t.a.v. de oliecrisis 1973/74, Folder 8003*, memo DES to Van der Stoel, no. 1006, 09/11/73.

21 NMFA, *Code 614.51, Folder* OESO *Oliecommissie, Volume V*, Van der Stoel to Brussels, no. 192, 14/11/74 en Van Schelle to NMFA, no. 395, 20/11/74.

22 HTK, *1973-1974*, pp. 1000-1004, en NMFA, BZ-*rapport Oliecrisis, Volume II*, p. 166.

23 NMFA, *Code 613.211.45, Box 38, Folder 440*, Rutten to Van der Stoel, Nota inzake Nederlandse voorzieningssituatie, 22/11/73.

24 AAZ, RVD-*Dossier Coördinatiegroep Voorlichting Energietekort*, Memorandum, November 1973.

25 AEZ, *Archive* PAZ, *Box 616, Folder 2*, Verslag 4de vergadering EZ werkgroep Oliecrisis, 21/11/73.

26 Maull refers to 're-exports from Rotterdam's refining centre, which fell to 39% of their normal level in the first half of December, but recovered to 90 per cent in January'. See: Maull, *1975*, p. 7.

27 HTK, *1973-1974, Bijlagen, 12724*, Nota over de beperking van de olie-aanvoer en de gevolgen daarvan, p. 2, 08/12/73.

28 NMFA, *Code 996.236.3, Folder 201*, memo DES to Van der Stoel via DGES en S, 23/11/73.

29 KMP, *Cabinet meetings*, minutes of Cabinet meeting, no. 6234, 30/11/73.

30 NNA, *2.02.05.02, Archive of the Council of Ministers, Inv. no. 1228*, Ontwerpnota "Energiepolitieke maatregelen samenhangende met de beperking van de aanvoer van ruwe olie naar Nederland", 10/12/73.

31 KMP, *Cabinet meetings*, minutes of Cabinet meeting, Council for Economic Affairs, 05/12/73.

32 AEZ, *Archive* RBA, *Box 20-0009, Folder 5*, Nota concept besluitvorming inzake aardgas, no. 45, 06/12/73.

33 AEZ, *Archive PAZ, Box 616, Folder 2*, Verslag 1ste vergadering EZ Werkgroep Oliecrisis, 07/11/73.
34 *KMP, Cabinet meetings*, minutes of Cabinet meeting, no. 6206, 09/11/73.
35 NNA, *2.02.05.02, Archive van de Raad van Ministers, inv. no. 1225*, Memorandum Van de Graaf to Den Uyl, no. 604, 15/11/73.
36 NNA, *2.02.05.02, Archive van de Raad van Ministers, inv. no. 1225*, Memorandum Van de Graaf to Den Uyl, no. 605, 15/11/73.
37 AEZ, *Archive RBA, Box 20-0009, Folder 5*, Derde Rapport Coördinatiegroep Maatregelen Oliecrisis, 22/11/73.
38 *KMP, Cabinet meetings*, minutes of Cabinet meeting, no. 6226, 23/11/73. These foreigners received a lot of attention at the NMFA. A flexible regime was preferred, as was the advice to Van der Stoel. Irritations in foreign countries had to be avoided. The interest of the Dutch trucking was paramount and should not be hindered. See: NMFA, *BZ-rapport Oliecrisis, Volume II*, p. 193.
39 AEZ, *Archive PAZ, Box 616, Folder 2*, Verslag 5de vergadering EZ werkgroep Oliecrisis, 28/11/73.
40 *KMP, Cabinet meetings*, minutes of Cabinet meeting, no. 6234, 30/11/73.
41 *HTK, 1973-1974, Bijlagen, 12724*, Nota over de beperking van de olie-aanvoer en de gevolgen daarvan, p. 2, 08/12/73.
42 Interview with Willemsen, February 1997 and letter by Willemsen to the authors, 26 May 1998.
43 NMFA, *BZ-rapport Oliecrisis, Volume II*, p. 197.
44 *HTK, 1973-1974, Bijlagen, 12723*, Memorie van Toelichting bij de Machtigingswet, 08/12/73.
45 Van Schendelen, *1979*, pp. 19-23.
46 Van Schendelen, *1979*, p. 11.
47 Van Schendelen, *1979*, pp. 12-14.
48 *HTK, Bijlagen, 12739*, Beleidsnota beperking gevolgen olieschaarste, 12/12/73.
49 *KMP, Cabinet meetings*, minutes of Cabinet meeting, no. 6272, 21/12/73.
50 *KMP, Cabinet meetings*, minutes of Cabinet meeting, no. 6272, 21/12/73.
51 *NRC Handelsblad*, 03/01/74.
52 *KMP, Cabinet meetings*, minutes of Cabinet meeting, no. 6274A, 04/01/74.
53 *De Volkskrant*, 05/01/74.
54 *Trouw*, 05/01/74.
55 *NRC Handelsblad*, 05/01/74.
56 AEZ, *Archive RBA, Box 20-0009, Folder 5*, Conclusies van het overleg Ministeriële Commissie Olieproblematiek, 10/01/74.
57 IISG, *Collection Den Uyl, Inv. no. 1110*, Van de Graaf to Den Uyl, 08/01/74.
58 *KMP, Cabinet meetings*, minutes of Cabinet meeting, no. 6285, 11/01/74.
59 AEZ, *Archive RBA, Box 20-0001, Folder 3*, Telex from BK gas BV to RBA, no. 603, 11/01/74.

60 NMFA, *BZ-rapport Oliecrisis, Volume II*, p. 200. In the beginning of January distribution was planned only in Sweden and Norway.

61 Interview with Rutten, January 1997. He referred to the behaviour of his own minister as typical Lubbers nervousness.

62 Interview with Rutten, January 1997. According to Rutten, Lubbers was not present because he did not dare to become responsible for the decision regarding distribution.

63 NMFA, *BZ-rapport Oliecrisis, Volume II*, p. 197.

64 *KMP, Cabinet meetings*, minutes of Cabinet meeting, no. 6274A, 04/01/74.

65 Interview with Helfrich, February 1997.

66 AEZ, *Archive RBA, Box 20-0009, Folder* 2, 'Rapport inzake de aardoliedistributie in de periode 19 oktober 1973 tot 28 februari 1974', Rijksbureau voor Aardolieprodukten, February 1974, pp. 12-32.

67 NMFA, *BZ-rapport Oliecrisis, Volume II*, p. 199.

68 AEZ, *Archive RBA, Box 20-0009, Folder* 2, Rapport inzake de aardoliedistributie in de periode 19 oktober 1973 tot 28 februari 1974, pp. 12-32.

69 AEZ, *Archive RBA, Box 20-0001, Folder* 4, Hustinx to Lubbers, no. 1360, 22/01/74.

70 *HTK, 1973-1974, Bijlagen, 12946*, Nota 'Kwantitatieve beperkingen in de olieaanvoer', p. 17, 04/06/74.

71 AMJ, *Dossier A 73/099, Folder energiecrisis, Volume I*, Nota Me. Mok to Van Agt ('extremely urgent'), 21/01/74.

72 NMFA, *Code 614.51, Folder 8003*, 9e rapport Coördinatiegroep Maatregelen Oliecrisis, 09/01/74 en AEZ, *Archive Bureau Secretaris Generaal, Box 290, Folder 8*, Nota EP to SG, 24/01/74.

73 NMFA, *BZ-rapport Oliecrisis, Volume II*, p. 200.

74 NNA, *2.02.05.02, Archive of the Council of Ministers, Inv. no. 1289*, Note Rutten and Van Agt to Den Uyl, 23/01/74.

75 *KMP, Cabinet meetings*, minutes of Cabinet meeting, no. 6304, 23/01/73.

76 *De Volkskrant*, 24/01/74.

77 *De Volkskrant*, 26/01/74.

78 NRC *Handelsblad*, 24/01/73.

79 NRC *Handelsblad*, 26/01/74.

80 AMJ, *Dossier A 73/099, Folder energiecrisis, Volume I*, Nota van het Hoofd van de Hoofdafdeling Staats- en Strafrecht L. Oranje to de SG, no. 13, 25/01/74.

81 AEZ, *Archive RBA, Box 20-0009, Folder 5*, Rapport Coördinatiegroep Maatregelen Oliecrisis, 07/02/74.

82 AEZ, *Archive BEB, no. 89-3321, Box 5649*, Overmars to Director Bilaterale Zaken, 'Mogelijkheden tot verbetering van het Nederlandse image in de Arabische landen; EEG-aspecten', no. 180, 01/02/74.

83 Interview with Helfrich, February 1997. Not only in The Netherlands was there a frantic search for extra capacity to store oil. Also in the USA companies procured and rented extra fuel tanks and thousands of train tank cars in order to store the oil. J.R. Scanlin, president of the *General American Transportation Corporation*, stated in December 1973: 'We had about 3,800 idle tank cars [available to lease] a year ago. Today we have 700'. See: Sherrill, *1983*, p. 195.

84 AMJ, *Dossier A 73/099, Folder energiecrisis, Volume II*, Note to Mr. Q.J.M Kramer, 26/03/74.

85 AEZ, *CAB, PAZ, Section DGE, Box 85-0947, Folder 2*, Rapport Aardolie-distributie 1974, distributiekring Amsterdam (vertrouwelijk), March 1974.

86 AEZ, *Archive RBA, Box 20-009, Folder 2*, 'Rapport inzake de aardoliedistributie in de periode 19 oktober 1973 tot 28 februari 1974', Rijksbureau voor Aardolieprodukten, February 1974, pp. 1-2.

87 See also: IISG, *Collection Den Uyl, Inv. no. 1110*, Note De Graaf to Den Uyl, 08/01/74.

88 See for this also: Mitchell, *1989*, p. 482.

89 *HTK, 1973-1974, Bijlagen, 12946*, 'Nota kwantitatieve beperkingen in de olieaanvoer', p. 16, 04/06/1974.

90 Wagner was warned for this. Not only Western intelligence services could intercept the Shell traffic but also Soviet and Warsaw Pact services could do the same. Shell was spurred to improve their communications security. However, this was not a unique Dutch capability. According to a former employee of the National Security Agency (NSA) this service intercepted and read the communications traffic of Esso, Exxon, American Shell and Mobil. Also the traffic of Japanese multinationals was an important target and in particular their confidential conversations with Yamani produced much valued information. Confidential interviews.

91 NMFA, *BZ-rapport Oliecrisis, Volume II*, p. 195.

92 NOVA interview with Westerterp, October 1993.

93 There were also problems with the Ministry of Defence. They considered themselves to have absolute priority and with success. Not a single military exercise was cancelled. Interview with Willemsen, February 1997.

Notes Chapter 7

1 Hoff, *1991*, pp. 108-109.

2 NMFA, *Code 996.236.3, OESO Oliecommissie, Volume III, Folder 201*, Van der Stoel to all embassies, no. 275, 30/11/73.

3 See for the text of all decisions taken: Fraser, *1980*, pp. 195-196 and Jaarboek, *1974*, p. 82.

4 NMFA, *Code 614.51, Folder 7806*, aide-mémoire, presented by the US ambassador to DGPZ, 22/12/73.
5 NMFA, *BZ-rapport Oliecrisis, Volume II*, p. 204.
6 NMFA, *Code 996.236.3, OESO Olie-Commissie, Volume IV, Folder 202*, Van der Stoel to London, no. 1, 03/01/74.
7 NMFA, *Code 996.236.3, OESO Olie-Commissie, Volume IV, Folder 202*, Van der Stoel to London, no. 3, 04/01/74.
8 NMFA, *Code 613.211.45, Box 38, Folder 440*, De Beus to NMFA, no. 496, 18/12/73.
9 NOVA interview with Van der Stoel, October 1993.
10 During a conversation with Van Elslande on 7 January, Van der Stoel and Lubbers agreed on this arrangement as long as it stayed exclusively within the Belgian-Dutch domain. See: NMFA, *Code 614.51, Folder 7814*, memo DAM/MO to DGPZ, no. 1/74, 03/01/74 and *Code 996.236.3, OESO Oliecommissie, Volume III, Folder 201*, Van Schelle to NMFA, no. 8, 07/01/74.
11 Yergin, *1991*, pp. 628-629 and Adelphi Paper no. 117, *1975*, p. 9.
12 NMFA, *Code 913.211.0, Folder 5702*, Van Oven to NMFA, no. 145, 28/12/73.
13 *International Herald Tribune*, 18/01/74. A study drafted by the *International Longshoremen's Association* stated later that '*oil shipments from the Arabian/Persian Gulf (including Iraq and Iran) in December 1973 were about 7.4 per cent below the September level*'. The US Department of Commerce later concluded that millions of barrels flowed into the United States form OAPEC countries during the embargo. Most of the oil (25.8 million barrels) came form Saudi Arabia. In short, the oil flow did not diminish considerably. Late January, the import of oil in Rotterdam was already at 80% of the pre-crisis situation. See: SIPRI, *1974*, p. 151.
14 NMFA, *Code 613.211.45, Box 38, Folder 440*, Derksen to NMFA, no. 3, 29/01/74.
15 NMFA, *Code 613.211.45, Box 38, Folder 440*, Derksen to NMFA, no. 7, 04/02/74.
16 NMFA, *BZ-rapport Oliecrisis, Volume II*, p. 207.
17 *NRC Handelsblad*, 14/01/74.
18 *Trouw*, 29/01/74.
19 NMFA, *Code 913.211.0, Folder 'Midden-Oosten, Olieboycot, Volume 3'*, Schorer to NMFA, no. 12, 27/01/74.
20 Isaacson, *1992*, pp. 542-550.
21 NMFA, *Code 913.211.0, Folder 5702*, Van der Stoel to Van Lynden, no. 12, 12/01/74 and Van Lynden to NMFA, no. 37, 14/01/74.
22 NMFA, *Code 996.236.3, OESO Oliecommissie, Volume III, Folder 201*, De Ranitz to NMFA, no. 49, 25/01/74.

23 NMFA, *Code 614.51, Folder 8003*, De Ranitz to NMFA, no. 6, 04/01/74.

24 NMFA, *Code 996.236.3*, OESO *Oliecommissie, Volume III, Folder 202*, Celer to Paris, no. 11, 24/01/74, and *Elsevier*, 2 February 1974.

25 NMFA, *Code 913.211.0, Folder 5706*, Schorer to NMFA, no. 3, 13/01/74.

26 NMFA, *Code 912.10, Folder 1939*, memo DAM to DGPZ, no. 15/74, 31/01/74.

27 NMFA, *Code 614.51, Folder 7835*, Madrid to NMFA, no. 106, 12/12/73.

28 International Economic Report, *1974*, p. 45.

29 Vernon, *1976*, pp. 113-127.

30 See:. *Het Parool*, 08/01/74 and 09/01/74.

31 NMFA, PA *Paris, Code 614.51* GS, *Box 1*, De Ranitz to NMFA, no. 767, 21/12/73.

32 SIPRI, *1974*, pp. 110-111.

33 NMFA, *Code 913.211.0, Folder 5702*, De Ranitz to NMFA, no. 55, 30/01/74.

34 SIPRI, *1974*, pp. 116-117.

35 NMFA, *Code 913.211.0, Folder 5702*, De Ranitz to NMFA, no. 60, 31/01/74.

36 NMFA, *Code 614.51, Folder 8083*, Gevers to NMFA, no. 5 & 33, 04/01/74 and 18/01/74. Political motives perhaps also played a role. Feisal wanted Heath to continue his pro-Arab policy and was hostile towards the pro-Israel attitude of the Labour Party led by Harold Wilson. Feisal even ordered that Britain should receive more Saudi oil than before the crisis broke out. See: Lieber, *1976*, p. 30.

37 NMFA, *Code 913.211.0, Folder 5702*, Gevers to NMFA, no. 31, 18/01/74; SIPRI, *1974*, p. 117, and Lieber, *1976*, p. 31.

38 Yergin, *1991*, p. 629.

39 NMFA, *Code 913.211.0, Folder 5706*, De Beus to NMFA, no. 142, 26/03/74.

40 NMFA, PA *Archive Bonn, Code 614.56, Box 19*, Derksen to NMFA, no. 30, 28/03/74.

41 SIPRI, *1974*, p. 116.

42 NMFA, *Code 913.211.0, Folder 5702*, Boon to NMFA, no. 27, 24/01/74.

43 SIPRI, *1974*, p. 116.

44 KMP, *Cabinet meetings*, minutes of Cabinet meeting, no. 6285, 11/01/74.

45 NMFA, *Code 913.211.0, Folder 5706*, Van der Stoel circular no. 40, 04/02/74.

46 KMP, *Cabinet meetings*, minutes of Cabinet meeting, no. 6315, 01/02/74.

47 AEZ, *Werkarchief BEB, 89-3321, Box 5649*, Note to Directeur Bilaterale Zaken, drafter Overmars. Onderwerp: 'Mogelijkheden tot verbetering van het Nederlandse image in de Arabische landen; EEG-aspecten', no. 180, 01/02/74.

48 KMP, *Cabinet meetings*, minutes of Cabinet meeting, no. 6315, 01/02/74.

49 Simonian, *1985*, pp. 212-213.

50 NMFA, *Code 996.236.3*, OESO *Oliecommissie, Volume IV, Folder 202*, memo DES to DGES, no. 83, 21/01/74.

51 Kissinger, *1982*, pp. 896-903.

52 NMFA, *Code 614.51, Folder 7803*, Memorandum Plan 'Kissinger/Energy Action Group' of DWH to DGPZ, 21/12/73.

53 NMFA, *Code 913.211.0, Folder 5702*, De Ranitz to NMFA, no. 11, 08/01/74.
54 NMFA, *Code 996.236.3, OESO Oliecommissie, Volume IV, Folder 202*, Letter Nixon to Den Uyl (secret), 09/01/74.
55 NMFA, *Code 996.236.3, OESO Oliecommissie, Volume IV, Folder 202*, memo Van Schaik, 08/01/74, en conclusies van de vergadering over multilateraal overleg inzake olie and energie, 09/01/74.
56 NMFA, *Code 996.236.3, OESO Oliecommissie, Volume IV, Folder 202*, memo Van Schaik, 08/01/74 and conclusies van de vergadering over multilateraal overleg inzake olie and energie, 09/01/74.
57 NMFA, *Code 614.51, Folder 8003*, memo DES to M, no. 39/74, 10/01/74.
58 NMFA, *BZ-rapport Oliecrisis, Volume II*, p. 215.
59 NMFA, *Code 996.40, Folder EG/Olie & energiecrisis, Volume 2, Folder 1848*, Van der Stoel, circular no. 1008, 10/01/74.
60 NMFA, *BZ-rapport Oliecrisis, Volume II*, p. 216.
61 Kissinger, *1982*, pp. 903-904.
62 NMFA, *Code 996.236.3, OESO Oliecommissie, Volume IV, Folder 202*, memo DES to DGES, no. 83, 21/01/74.
63 NMFA, *Code 996.236.3, OESO Oliecommissie, Volume IV, Folder 202*, Van der Stoel to Washington, no. 53, 01/02/74.
64 *KMP, Cabinet meetings*, minutes of Cabinet meeting, no. 6315, 01/02/74.
65 See for this also Chapter 8.
66 NMFA, *Code 996.236.3, OESO Oliecommissie, Volume IV, Folder 202*, Sassen to NMFA, no. 22, 24/01/74.
67 NMFA, *Code 913.211.0, Folder 5702*, De Ranitz to NMFA, no. 29, 17/01/74.
68 *Trouw*, 02/02/74.
69 Grünfeld, *1991*, p. 86; Simonian, *1985*, pp. 230-231 and Mendershausen, *1976*, p. 72.
70 NMFA, *PA Djedda, Code 614.0 GS, Box 39, Folder Internationaal overleg inzake grondstoffen*, Van der Stoel, circular no. 49, 06/02/74.
71 NMFA, *Code 996.236.3, OESO Oliecommissie, Volume IV, Folder 202*, Kaufmann to NMFA, no. 12, 24/01/74.
72 NMFA, *BZ-rapport Oliecrisis, Volume II*, p. 221.
73 NMFA, *BZ-rapport Oliecrisis, Volume II*, p. 222.
74 Kissinger, *1982*, pp. 899-890.
75 NMFA, *BZ-rapport Oliecrisis, Volume II*, p. 223 and Kissinger, *1982*, pp. 899-890.
76 Jobert, *1974*, pp. 284-288.
77 NMFA, *Code 614.51, Folder 7803*, De Roos to NMFA, no. 276/64, 11/02/74.
78 Joustra & van Venetië, *1989*, pp. 132-133.
79 Kissinger, *1982*, pp. 910-912.

80 Interview with Van der Stoel, January 1997.

81 NMFA, *Code 614.51, Folder 7803*, Van Lynden to NMFA, no. 380/738, 11/02/74.

82 Goldsborough, *1974*, p. 545.

83 Simonian, *1985*, pp. 232-241; Jobert, *1976*, pp. 378-383 and Yergin, *1991*, p. 630.

84 'Jobert, personifying to the Germans the delusions in French policy, was the butt of Schmidt's anger.' See: Bark & Gress, *1993*, pp. 297-300.

85 Schmidt, *1987*, pp. 201-202. Schmidt pleaded already on 4 November 1973 in a letter to Kissinger for a kind of consumers' facade against the OPEC.

86 SIPRI, *1974*, pp. 33 and 124-131.

87 NMFA, *Code 999.214, Folder 684*, Van der Stoel circular no. 58, 15/01/74; Kissinger, *1982*, pp. 912-925 and Lieber, *1976*, pp. 22-25.

88 Costigliola, *1992*, p. 180.

89 *KMP, Cabinet meetings*, minutes of Cabinet meeting, no. 6332, 15/02/74.

90 NMFA, *Code 614.51, Folder 7803*, Van der Stoel circular no. 58, 15/02/74. NNA, *2.02.05.02, Archive of the Council of Ministers, Inv. no. 1309*, Note by DES to the Council, no. DES 76451, 29/03/74.

91 Simonian, *1985*, pp. 225-228.

Notes Chapter 8

1 SIPRI, *1974*, p. 29 and Kalb & Kalb, *1975*, pp. 588-589.

2 Hoff's conclusion that Algeria was opposed to the lifting of the embargo is not correct. Hoff, *1991*, p. 108.

3 NMFA, *BZ-rapport Oliecrisis, Volume II*, p. 209.

4 NMFA, *Code 613.211.45, Folder 441*, Van der Stoel to Washington, no. 43, 30/01/74.

5 NMFA, *Code 613.211.45, Folder 441*, Van Lynden to NMFA, no. 105, 01/02/74.

6 NMFA, *Code 913.211.0, Folder 5706*, undated draft. It is unclear if this text was transmitted in the end.

7 NMFA, *Code 913.211.0, Folder 5706*, Derksen to NMFA, no. 10, 09/02/74.

8 NMFA, *Code 913.211.0, Folder 5706*, Schorer to NMFA, no. 21, 17/02/74.

9 NMFA, *BZ-rapport Oliecrisis, Volume II*, p. 244.

10 NMFA, *Code 912.1 GS, Folder 1026*, Derksen to NMFA, no. 15, 28/02/74.

11 NMFA, *BZ-rapport Oliecrisis, Volume II*, p. 245.

12 Yergin, *1991*, p. 631; Daoudi & Dajani, *1985*, pp. 143-144 and Mitterand, *1982*, pp. 106-111.

13 Kissinger, *1982*, pp. 946-952. Also the embargo against Rhodesia, Portugal and South Africa was continued.

14 NMFA, *Code 913.211.0, Folder 5706*, Van der Stoel circular no. 1048, 22/03/74.
15 NMFA, *Code 614.51, Folder 8008*, Van der Stoel to Copenhagen, no. 8, 19/03/74.
16 NRC *Handelsblad*, 19/03/74.
17 NMFA, *BZ-rapport Oliecrisis, Volume II*, p. 246.
18 NMFA, *PA Algiers, Code Folder 614.56, Box 19*, Bot to NMFA, no. 29, 20/03/74.
19 NMFA, *PA Algiers, Code 614.56, Box 19*, Van Hoeve to NMFA, no. 33, 24/03/74 and Bentinck to NMFA, no. 26, 23/03/74.
20 NMFA, *Code 913.211.0, Folder 5706*, De Ranitz to NMFA, no. 195, 25/03/74.
21 NMFA, *Code 913.211.0, Folder 5706*, De Ranitz to NMFA, no. 196, 25/03/74.
22 NMFA, *Code 913.211.0, Folder 5706*, Derksen to NMFA, no. 30, 28/03/74. Akins would have described the continuation of the embargo as '*petty*'. See: Van der Stoel to Djedda, no. 19, 28/03/74.
23 NMFA, *Postarchief Algiers, Code 614.56, Box 19*, Kaufmann to NMFA, no. 185, no. 190 and no. 191, 11/04/74. Also: Beelaerts to NMFA, no. 39, 13/04/74 and Terlingen & Roskam, *1997*, pp. 64-65.
24 NMFA, *Code 999.214, Folder 684*, Fack to NMFA, no. 24, 17/01/74.
25 NMFA, *Code 999.214, Folder 684*, Fack to NMFA, no. 34, 23/01/74.
26 NMFA, *Code 999.214, Folder 684*, Fack to NMFA, no. 37, 25/01/74.
27 NMFA, *Code 996.236, Folder 202*, Celer to PVEG, Brussels, no. 8, 22/01/74.
28 NMFA, *BZ-rapport Oliecrisis, Volume II*, p. 219.
29 NMFA, *Code 999.214, Folder 684*, Pronk to New York, no. 25, 24/01/74.
30 KMP, *Cabinet meetings*, minutes of the Cabinet meeting, no. 6315, 01/01/74.
31 NMFA, *Code 913.211.0, Folder 5702*, Van der Stoel to Algiers, no. 7, 08/02/74.
32 NMFA, *Code 912.1, Folder 2089*, Pronk to Algiers, no. 6, 08/02/74.
33 ASD, *FOIA 8903648*, State Department memorandum, 11/04/74.
34 NMFA, *BZ-rapport Oliecrisis, Volume II*, pp. 235-236.
35 NMFA, *BZ-rapport Oliecrisis, Volume II*, pp. 237-239.
36 NMFA, *Code 913.211.0, Folder 5706*, Van der Stoel to Damascus, no. 16, 20/03/74.
37 KMP, *Cabinet meetings*, minutes of Cabinet meeting, no. 6365, 08/03/74.
38 NMFA, *BZ-rapport Oliecrisis, Volume II*, p. 240.
39 NMFA, *Code 996.0 EEG, Folder 594, Volume 17, EEG-VS*, Van Aerssen to Van der Stoel, no. 381 142, 06/03/74; Kissinger, *1982*, pp. 927-931 and Simonian, *1985*, pp. 241-242.
40 NMFA, *BZ-rapport Oliecrisis, Volume II*, p. 240.
41 NA, *RG59, Kissinger Staff Meetings 1973-1977, Entry 5177, Box 2*, 05/03/74 and *Box 3*, 22/03/74.
42 NMFA, *BZ-rapport Oliecrisis, Volume II*, p. 241.
43 NMFA, *PA London, GS, Code 912.1, Box 44*, Van der Stoel circular no. 86,

11/03/74. About Callaghan's 'warm feelings' for the United States: Callaghan, *1987*, pp. 358-359.
44 NA, *RG59, Kissinger Staff Meetings 1973-1977, Entry 5177, Box 3*, 22/03/74.
45 NMFA, *Code 996.0 EEG, Folder 594, Volume 17, EEG-VS*, Van der Stoel to Den Uyl, no. 69046, 22/03/74.
46 Marshall, *1997*, p. 91 and Simonian, *1985*, p. 244.
47 NMFA, *Code 996.0 EEG, Folder 594, Volume 17, EEG-VS*, Tammenons Bakker (van M) to NMFA, no. 86, 25/04/74.
48 NMFA, *BZ-rapport Oliecrisis, Volume II*, pp. 242-243.
49 Agulhon, *1993*, pp. 439 and 449 and Costigliola, *1992*, pp. 180-182.
50 Marshall, *1997*, pp. 92-93.
51 AEZ, *Archive PAZ, Box 616, Folder 2*, Verslag 13de vergadering EZ Werkgroep Oliecrisis, 30/01/74.
52 NNA, *2.02.05.02, Archive of the Council of Ministers, Inv. no. 1289*, minutes of the Cabinet meeting, no. 6354, 01/03/74.
53 *Keesings Historisch Archief*, 28 June 1974, p. 406.
54 AEZ, *Archive Bureau Secretaris-Generaal, Box 294, Folder 11*, Directie AEP to SG, no. 303/74/EP, 19/04/74.
55 *Keesings Historisch Archief*, 15 February 1974, p. 101.
56 NMFA, *Code 913.211.0, Folder 5706*, De Beus to NMFA, no. 148, 28/03/74.
57 NMFA, *BZ-rapport Oliecrisis, Volume II*, pp. 244-248.
58 NMFA, *Code 913.211.0, Folder 5706*, memo DAM to M, no. 40 plus annexes, 28/03/74.
59 NMFA, *BZ-rapport Oliecrisis, Volume II*, pp. 249-250.
60 NMFA, *Code 913.211.0, Folder 5706*, Van der Stoel to Bonn, no. 97, 02/04/74.
61 AEZ, *Archive DG BEB, no. 89-1227, Box 3283, Folder 7*, Note by A. Houtman, 15/03/74.
62 Our reconstruction is based on: Ministerie van Buitenlandse Zaken, *1974*, passim.
63 *KMP, Cabinet meetings*, minutes of Cabinet meeting, no. 6421, 19/04/74.
64 NMFA, *PA Djedda, GS, Folder 614.0*, Van Lynden to NMFA, no. 313, 15/04/74.
65 *KMP, Cabinet meetings*, minutes of Cabinet meeting, no. 6441, 03/05/74.
66 NMFA, *BZ-rapport Oliecrisis, Volume II*, p. 233.
67 NNA, *2.02.05.02, Archive of the Council of Ministers, Inv. no. 1289*, minutes of the Cabinet meeting, no. 6410, 11/04/74.
68 NMFA, *BZ-rapport Oliecrisis, Volume II*, p. 253.
69 NMFA, *Code 913.211.0, Folder 5706*, Derksen to NMFA, no. 31, 28/03/74.
70 NMFA, *Code 913.211.0, Folder 5706*, Van Schelle to NMFA, no. 135, 29/03/74.
71 NMFA, *BZ-rapport Oliecrisis, Volume II*, p. 254.
72 Quandt, *1977*, pp. 231-245.

73 NMFA, *Code 912.2, Folder 1013*, Derksen to NMFA, no. 48, 01/06/74.
74 NMFA, *Code 913.211.0, Folder 5706*, Van Hoeve to NMFA, no. 63, 03/06/74, and *International Herald Tribune*, 02/06/74.
75 NMFA, *Code 913.211.0, Folder 5706*, Reinink to NMFA, no. 77, 06/06/74.
76 NMFA, *Code 913.211.0, Folder 5706*, De Ranitz to NMFA, no. 316, 07/06/74, and *Le Monde*, 04/06/74.
77 Interview with Van der Stoel, January 1997.
78 NMFA, *Code 913.211.0, Folder 5706*, memo DAM to M, no. 85, 07/06/74.
79 *KMP, Cabinet meetings*, minutes of the Cabinet meeting, no. 6485, 07/06/74.
80 NMFA, *Code 913.211.0, Folder 5706*, Derksen to NMFA, no. 50, 13/06/73, and Celer to Djedda, no. 31, 17/06/74.
81 NMFA, *BZ-rapport Oliecrisis, Volume II*, pp. 255-256.
82 NMFA, *Code 913.211.0, Folder 5706*, De Ranitz to NMFA, no. 339, 18/06/74.
83 NMFA, *Code 913.211.0, Folder 5706*, Vixseboxse to NMFA, no. 95, 24/06/74.
84 NMFA, *Code 913.211.0, Folder 5706*, Schaapveld to NMFA, no. 272, 20/06/74.
85 NMFA, *Code 912.1, Folder 2274*, Memo DGPZ to M, 28/06/74.
86 NRC *Handelsblad*, 10/07/74.
87 Daoudi & Dajani, *1985*, p. 155, and ASD, *FOIA 8903648*, State Department Briefing Paper, May 1974.
88 NNA, *2.02.05.02, Archive of the Council of Ministers, Inv. no. 1327*, Note DES to the Council, no. DES 129754, 05/07/74.
89 NNA, *2.02.05.02, Archive of the Council of Ministers, Inv. no. 1327*, Note Van de Graaf to Den Uyl, no. 315, 13/06/74.
90 Lieber, *1976*, p. 38 and Jaarboek, *1974*, pp. 82-86.
91 NNA, *2.02.05.02, Archive of the Council of Ministers, Inv. no. 1289*, minutes of the Cabinet meeting, no. 6563, 12/07/74.

Notes Conclusion

1 Daoudi and Dajani, *1985*, p. 173-174.
2 Vernon, *1976*, pp. 101-102.
3 Voorhoeve, *1979*, p. 245. The embargo 'showed that Dutch foreign policy could be relatively independent from economic self-interest'. Voorhoeve has to acknowledge, however, that Dutch national security policy became more pro-Arab after 1974.
4 NMFA, *BZ-rapport Oliecrisis, Volume I*, p. 108.
5 According to former advisor of Prime Minister Den Uyl, T. van de Graaf. Interview with Van de Graaf, January 1997.
6 Vernon, *1976*, pp. 163-164.

7 According to former minister Chr. van der Klaauw. Interview with Van der Klaauw, September 1993.

8 *De Staatscourant*, 17/04/74.

9 Tweede-kamerfractie Partij van de Arbeid, 'Het onderzoek naar de rol van de oliemaatschappijen in de oliecrisis van 1973-1974', pp. 14-15.

10 Interview with Rutten, January 1997.

11 Interview with Willemsen, February 1997.

12 Interview with Van de Graaf, January 1997.

13 Voorhoeve, *1979*, p. 246.

Archival Records

A large number of records and collections of documents were consulted for the study. Private organizations and individuals also made documents available. The aim of this survey is to provide insight into the archives which were consulted for the purposes of this study.

Netherlands Ministry of Foreign Affairs archives

1 Secret Archives
2 Departmental Archives
3 Archives various diplomatic posts

The Netherlands Ministry of Defence archives

1 Department of the Under-Secretary of Defence
2 Secretary-General's Department
3 Defence Staff
4 Directorate-General for Equipment
5 Military Intelligence Service (MIS), Central Organization

Netherlands National Archives, The Hague

1 Minutes of Cabinet Meetings
2 National Defence Council
3 Collection J.H. van Royen
4 Collection H.J. de Koster

Archives City of Amsterdam

Newspaper Archives

Archives City of Rotterdam
Central Archives

Ministry of Home Affairs, The Hague
Collection of the Dutch National Security Service (BVD) in Leidschendam

International Institute of Social History, Amsterdam
Collection E. van Thijn
Collection J.M. den Uyl

General Affairs, The Hague
1 The archive of the Prime Minister's office (KMP)
2 Minutes of the Ministerial Council for European Affairs
3 Minutes of the Ministerial Council for Economic Affairs
4 The archive of the Committee of the United Intelligence Services in the Netherlands (CVIN)
5 The archive of the Intelligence Coordinator

Ministry of Economic Affairs, The Hague
1 Secretary-General's Department
2 Archives of the Minister and Under-Secretary
3 Council of Director-Generals
4 Directorate for Foreign Economic Relations
5 National Bureau for Oil Products
6 Directorate-General for General Economic Policy
7 Directorate-General for Energy

Ministry of Finance, The Hague
General Treasury Archives
IMF Archives

Ministry of Justice, The Hague
Files A 73/099, Folder I & II, 1973-1976
File A 73/399

Ministry of Transport and Water Management, The Hague
Departmental Archives

National Archives, College Park, Maryland
RG 59 State Department diplomatic records

RG 165 Modern Military Field Branch
RG 273 National Security Council
RG 319 US Army Intelligence, G-2
RG 341 US Air Force Intelligence

National Archives, Alexandra, Virginia <C>
1 Richard Nixon Presidential Materials Project

US Department of State, Washington D.C.
FOIA requests

Washington National Records Center, Suitland, Maryland
RG 84 Diplomatic Posts (The Hague)

Correspondence and interviews were held with the following persons. Their position in 1973-1974 is indicated.

Agt, A.A.M. van	Minister of Justice
Bar On, C.	Israeli Ambassador in The Hague
Bergh, H. van den	Member of parliament for the PvdA
Gould Jr., Kingdon	US Ambassador in The Netherlands
Graaf, T. van de	Advisor of the Prime Minister to the Prime Minister's Office
Hek, A. van der	Member of parliament for the PvdA
Helfrich, A.C.	Commercial Director Shell Netherlands
Heuven, M. van	US diplomat in The Hague
Kaufman, J.	Permanent Representative of The Netherlands at the OESO
Klaauw, Chr. van der	Deputy-Permanent Representative of The Netherlands at the Mission to the United Nations
Kruimink, F.E.	Coordinator of The Netherlands intelligence community
Lubbers, R.F.M.	Minister of Economic Affairs
Meines, T.	Brigadier-general, Head logistics at the Ministery of Defence
Molenaar, H.	Director Rotterdam Harbor
Oskam, J.	Representative of the Independent Oil Traders in Rotterdam
Peijnenburg, G.H.J.	Secretary-General at the Ministery of Defence
Pronk, J.P.	Minister for Development Cooperation
Rabbani, M.	Consul of Kuwait in The Hague
Ranitz, J.A. de	Netherlands Ambassador in Paris
Riddle, A.W.	Military Air attaché at the US Embassy in The Hague
Rising, L.E.	Military attaché at the US Embassy in The Hague

Rutten, F.W.	Secretary-General at the Ministry of Economic Affairs and Chairman of Co-ordination Committee for the Oil crisis
Schaik, R.J. van	Chief Directorate-General European Cooperation at the Ministry of Foreign Affairs
Schiff, E.L.C.	Secretary-General at the Ministry of Foreign Affairs
Stemerdink, A.	Under-Secretary of Defence
Stoel, M. van der	Minister of Foreign Affairs
Swift, Carleton B.	CIA Station chief in The Hague
Tanguy, Charles Reed	US Counselor at the US Embassy in The Hague
Tjeenk Willink, H.D.	Advisor at the Prime Minister's office
Thijn, E. van	Leader of the PvdA in parliament
Vredeling, H.	Minister of Defence
Willemsen, W.Q.J.	Secretary of the Oil Crisis Working group
Wagner, G.A.	Chief executive of Royal Dutch Shell
Westerterp, T.	Minister of Transport and Water Management
Yamani, Ahmed Zaki	Oil Minister of Saudi Arabia

And various officials of The Netherlands intelligence community and officials working for foreign intelligence and security services who want to remain anonymous.

List of Acronyms and Terms

AAW	Archives of the Netherlands Embassy, Washington
AAZ	Archives of General Affairs (Cabinet's Office)
AEZ	Archives of the Ministry of Economic Affairs
AMD	Archives of the Ministry of Defence
AMF	Archives of the Ministry of Finance
AMJ	Archives of the Ministry of Justice
ARA	Netherlands National Archives
ARP	Anti-Revolutionary Party
ASD	Archives of the US State Department
AZ	Ministry of General Affairs
BEB	Directorate-General for the Foreign Economic Relations of the Ministry of Economic Affairs
BP	British Petroleum
BVD	Internal Security Service
BZ	Ministry of Foreign Affairs
CEC	Central-Economic Commission
CHU	Christian Historical Union
CIA	Central Intelligence Agency
CMO	Co-ordination Group Measure against Oilcrisis
COS	Chief of Station of the CIA
CPB	Central Planning Bureau
CPN	Communist Party Netherlands
CVIN	Committee on the United Intelligence Services in the Netherlands
DAM	Department for Africa and Middle-East of the Ministry of Foreign Affairs
DES	Department for Economic Cooperation of the Ministry of Foreign Affairs
DGES	Directorate-General for European Cooperation of the Ministry of Foreign Affairs

DGE	Directorate-General for Energy of the Ministry of Economic Affairs
DGI	Directorate-General Industry of the Ministry of Economic Affairs
DGIS	Directorate-General for International Cooperation of the Ministry of Foreign Affairs
DGPA	Directorate-General for Political Affairs of the Ministry of Foreign Affairs
DIE	Department European Integration of the Ministry of Foreign Affairs
DIO	Department for International Organizations of the Ministry of Foreign Affairs
DGV	Directorate-General for Traffic of the Ministry of Transport and Water Management
DM	Deutsche Mark
DOC	Defence Oil Center
ECD	Economic Control Service
ECG	Energy Co-ordinating Group
ECOSOC	Economic and Social Council of the UN
EEC	European Economic Community
EC	European Community
EMU	European Monetary Union
EPC	European Political Cooperation
EZ	Ministry of Economic Affairs
FEO	Federal Energy Office
GATT	General Agreement on Tariffs and Trade
HTK	Parliamentary Proceedings
IAEA	International Atomic Energy Agency
IDB	Dutch Foreign Intelligence Service
IEA	International Energy Agency
IEP	International Energy Programme
IMF	International Monetary Fund
KMP	Prime Minister's Office
KVP	Catholic People's Party
LPG	Liquefied Petrol Gas
M	Minister of Foreign Affairs
MAAG	Military Assistance Advisory Group
MDAP	Mutual Defense Assistance Program
MID	Military Intelligence Service
NATO	North Atlantic Treaty Organization
NKV	Dutch Catholic Trade Union
NMFA	Netherlands Ministry of Foreign Affairs
NS	Dutch Railways
NVV	Dutch Union of Trade Unions

OAPEC	Organization of Arab Petroleum Exporting Countries
OFE	US Office of Fuel and Energy
OPEC	Organization of Petroleum Exporting Countries
PA	Archives diplomatic post
PLAN	Department for Planning at the Ministry of Foreign Affairs
PLO	Palestine Liberation Organization
PPR	Radical Political Party
PSP	Pacifist Cooperation Socialist Party
PvdA	Labour Party
PV	Permanent Representative
R	Minister of Development Aid
PV	Council for Economic Affairs
S	Secretary-General of the Ministry of Foreign Affairs
SALT	Strategic Arms Limitation Talks
SER	Social Economic Council
T	Under-Secretary of European Affairs at the Ministry of Foreign Affairs
UNEF	United Nations Emergency Force
VVD	People's Party for Freedom and Democracy
WKC	Mathematical Center
WSAG	Washington Special Action Group
Z	Under-Secretary of Disarmament Affairs at the Ministry of Foreign Affairs

Bibliography

Aarts, P., *Je hebt olie en je wil wat. Visies op de eerste oliecrisis*, Amsterdam: Mera, 1987

Aarts, P., 'Olie als wapen', in: R. Doom en G. Nonneman (ed.), *Het Midden-Oosten hertekend*, Brussel: VUBPress, 1996, pp. 71-96

Abir, M., *Oil, Power and Politics. Conflict in Arabia, the Red Sea and the Gulf*, London: Frank Cass, 1974

Agulhon, M., *The French Republic 1879-1992*, Oxford: Blackwell, 1993

Aker, F., *October 1973. The Arab-Israeli War*, Hamden: Archon, 1985

Akins, J., 'The Oil Crisis: This Time The Wolf is Here', in: *Foreign Affairs*, vol. 51, no. 3 (April 1973), pp. 462-490

Anderson, J. & J. Boyd, *Fiasco*, New York: Times Books, 1983

Andrew, C., *For the President's Eyes Only. Secret Intelligence and the American Presidency from Washington to Bush*, London: HarperCollins, 1995

Bailey, S.D., *The Making of Resolution 242*, Dordrecht: Nijhoff, 1985

Balance, R. & S. Sinclair, *Collapse and Survival*, London: Allen and Unwin, 1983

Bar-Siman-Tov, Y., *Israel, the Superpowers and the War in the Middle East*, New York: Praeger, 1987

Bark, D.L. & D.R. Gress, *Democracy and Its Discontents 1963-1991*, Oxford: Blackwell, 1993

Black, C.E., et al, *Rebirth. A History of Europe Since World War II*, Boulder: Westview Press, 1992

Black, I. & B. Morris, *Israel's Secret Wars. A History of Israel's Intelligence Services*, London: Futuru, 1991

Bleich, A., *Een Partij in de Tijd. Veertig jaar Partij van de Arbeid 1946-1986*, Amsterdam: Arbeiderspers, 1986

Boer, J. & J. Oude Lohuis, *Olieslaaf. Met een goed energiebeleid naar onafhankelijkheid*, Amsterdam: Ekologische uitgeverij, 1983

Boon, H.N., *Afscheidsaudiëntie. Tien studies over de diplomatieke praktijk*, Rotterdam: Donker, 1976
Bouwmeester, P., *U weet het niet van mij. Politiek-journalistieke herinneringen*, Den Haag: SDU, 1991
Boyne, W.J., *The Two O'Clock War. The 1973 Yom Kippur Conflict and the Airlift that Saved Israel*, New York, 2002
Brandt, W., *Herinneringen*, Utrecht/Antwerpen: Veen, 1990
Brecher, M., *Decisions in Crisis. Israel 1967 and 1973*, London: Frank Cass, 1980
Breedveld, W. & J. Jansen van Galen, *Gaius. De onverstoorbare gang van W.F. de Gaay Fortman*, Utrecht: Scheffers, 1996
Brogan, P., *The Fighting Never Stopped*, New York: Vintage, 1990
Bromley, S., *American Hegemony and World Oil: The Industry, the State System and the World Economy*, Cambridge: Polity Press, 1991
Buchan, D., *Europe. The Strange Superpower*, Aldershot: Dartmouth, 1993

Callaghan, J., *Time & Change*, Glasgow: Collins, 1987
Campbell, J., *Edward Heath. A Biography*, London: Cape, 1993
Church, F., 'The Impotence Of Oil Companies', in: *Foreign Policy*, no. 27 (Summer 1977), pp. 27-51
Costigliola, F., *France and the United States. The Cold Alliance Since World War II*, New York: Twayne, 1992

Daltrop, A., *Politics and the European Community*, Harlow: Longman, 1985
Daoudi, M.S. & M.S. Dajani, *Economic Sanctions. Ideals and Experience*, London: Routledge, 1983
Daoudi, M.S. & M.S. Dajani, *Economic Diplomacy. Embargo Leverage and World Politics*, Boulder: Westview Press, 1985
Doran, C.F., *Myth, Oil, and Politics: Introduction to the Political Economy of Petroleum*, New York: Free Press, 1977

Ehmke, H., *Mittendrin. Von der Grossen Koalition zur Deutschen Einheit*, Hamburg: Rowohlt, 1996
Enders, T.H., 'OPEC and the Industrial Countries: The Next Ten Years', in: *Foreign Affairs*, vol. 53, no. 4 (July 1975), pp. 625-637

Fack, R., *Gedane Zaken. Diplomatieke Herinneringen*, Amsterdam: Sijthoff, 1984
Fraser, T.G., *The Middle East, 1914-1979*, London: Edward Arnold, 1980

Gaddis, J.L., *Strategies of Containment*, Oxford: OUP, 1982
Gann, L.H. & P. Duignan, *Contemporary Europe and the Atlantic Alliance. A Po-*

litical History, Oxford: Blackwell, 1998
Garthoff, R.L., *Détente and Confrontation. American-Soviet Relations from Nixon to Reagan*, Washington: Brookings, 1985
Geschiedenis van het Moderne Nederland. Politiek, economische en sociale ontwikkelingen, Houten: De Haan, 1988
Ginkel, J. van, *Het Westen en de oliecrisis 1973-1974*, 's-Gravenhage: NIVV-reeks 13, 1978
Golan, M., *The Secret Conversations of Henry Kissinger. Step-by-Step Diplomacy in the Middle East*, New York: Quadrangle, 1976
Goldsborough, J., 'France, the European Crisis and the Alliance', in: *Foreign Affairs*, vol. 52, no. 3 (April 1974), pp. 538-555
Goldstein, D.J. (ed.), *Energy and National Security*, Washington DC: National Defense University Press, 1982
Graaff, B. de & C. Wiebes, *Villa Maarheeze. De geschiedenis van de Inlichtingendienst Buitenland, 1946-1994*, Den Haag: Sdu, 1998
Grosser, A., *Les Occidentaux. Les pays d'Europeet les États-Unis depuis la guerre*, Parijs: Fayard, 1978
Grosser, A., *The Western Alliance. European-American Relations Since 1945*, New York: Vintage, 1982
Grünfeld, F., *Nederland en het Nabije Oosten: de Nederlandse rol in de internationale politiek ten aanzien van het Arabisch-Israëlische conflict 1973-1982*, Maastricht, 1991

Hanrieder, W.F. & G.P. Auton, *The Foreign Policies of West Germany, France and Britain*, New Jersey: Prentice Hall, 1980
Healey, D., *The Time of My Life*, London: Penguin, 1990
Heikal, M., *The Road to Ramadan*, London: Collins, 1975
Heikal, M., *The Sphinx and the Commissar. The Rise and Fall of Soviet Influence in the Middle East*, New York: Harper & Row, 1978
Hellema, D., *Negentienzesenvijftig. De Nederlandse houding ten aanzien van de Hongaarse revolutie en de Suezcrisis*, Amsterdam: Uitgeverij Jan Mets, 1990
Hellema, D., *Buitenlandse politiek van Nederland*, Utrecht: Aula, 1995
Hoff, R., *Het Midden-Oosten. Een politieke geschiedenis*, Utrecht: Aula, 1991
Houten, A. van den & M. Kopuit, *Wij staan achter Israël. De evenwichtige politiek van Nederland*, Amstelveen: Amphora, 1981
Hunter, R.E. (ed.), 'The Energy Crisis and U.S. Foreign Policy', in: *Headline Series*, no. 216, June 1973
Hyland, W.G., *Mortal Rivals. Understanding the Pattern of Soviet-American Relations*, New York: Touchstone, 1987

International Economic Report of the President, Washington: GPO, 1974
Isaacson, W., *Kissinger. A Biography*, New York: Simon & Schuster, 1992

Jaarboek van het Departement van Buitenlandse Zaken 1973-1974, 's-Gravenhage: Staatsuitgeverij, 1974
Jansen, M. & J.K. De Vree, *The Ordeal of Unity. The politics of European integration 1945-1985*, Bilthoven: Prime Press, 1985
Jenkins, R. (ed.), *Britain and the EEC*, London: Macmillan, 1983
Jobert, M., *Mémoires d'Avenir*, Paris: Bernard Grasset, 1974
Jobert, M., *L'autre regard*, Paris: Bernard Grasset, 1976
Jørgensen, A., *Bølgegang. Fra mine dagbøker 1972-1975. Bind 1*, København: Fremad, 1989
Joustra, A. & E. van Venetië, *Ruud Lubbers. Manager in de politiek*, Baarn: Anthos, 1989

Kalb, M. & B. Kalb, *Kissinger*, New York: Dell, 1975
Kissinger, H., *The White House Years*, London: Weidenfeld & Nicolson, 1979
Kissinger, H., *Years of Upheaval*, Boston/Toronto: Little, Brown and Company, 1982
Klaauw, C.A. van der, *Een diplomatenleven. Memoires*, Amsterdam: Bakker, 1995
Knoester, A., *Economische politiek in Nederland*, Leiden: Stenfert Kroese, 1989
Kohl, W.L. (ed.), *After the Second Oil Crisis. Energy Policies in Europe, America, and Japan*, Lexington: Heath and Company, 1982
Kossmann, E.H., *De Lage Landen 1780-1980. Twee eeuwen Nederland en België, Deel II 1914-1980*, Amsterdam: Elsevier, 1986
Krapels, E.N., *Oil and Security. Problems and Prospects of Importing Countries*, London: Adelphi Paper no. 136, 1977

Lebow, R.N. & J. Gross Stein, *We All Lost the Cold War*, Princeton: Princeton University Press, 1994
Lennep, E. van, *In de wereldeconomie. Herinneringen van een internationale Nederlander*, Leiden: Stenfert Kroese, 1991
Levy, W.J., 'World Oil Cooperation or International Chaos', in: *Foreign Affairs*, vol. 52, no. 3 (April 1974), pp. 690-713
Lieber, R.J., *Oil and the Middle East War: Europe in the Energy Crisis*, Harvard: Harvard Studies in International Affairs, 1976
Long, D.E., *Saudi Arabia*, Beverly Hills/London: Sage, 1976
Louis, W.M.R. & H. Bull, *The Special Relationship. Anglo-American Relations since 1945*, Oxford: Clarendon Press, 1989

Mancke, R.B., *Squeaking By. U.S. Energy Policy Since the Embargo*, New York: Columbia University Press, 1976

Marshall, B., *Willy Brandt. A Political Biography*, London: MacMillan, 1997

Maull, H., *Oil and Influence. The Oil Weapon Examined*, London: Adelphi Paper no. 117, 1975

Mendershausen, H., *Coping with the Oil Crisis. French and German Experiences*, Baltimore: John Hopkins University Press, 1976

Ministerie van Buitenlandse Zaken, *Verslag over de Zesde Bijzondere Zitting van de Algemene Vergadering der Verenigde Naties inzake Grondstoffen en Ontwikkeling, New York (9 april-2 mei 1974)*, no. 106, 's-Gravenhage, Staatsuitgeverij, 1974

Mitchell, B.R., *International Historical Statistics Europe 1750-1988*, Londen: Stockton Press, 1989

Mitterrand, F., *The Wheat and the Chaff. The Personal Diaries of the President of France 1971-1978*, London: Weidenfeld & Nicolson, 1982

Moore, J.N. (ed.), *The Arab-Israeli Conflict Readings and Documents*, Princeton: Princeton University Press, 1977

Nathan, J.A. & J.K. Oliver, *United States Foreign Policy and World Order*, Boston: Little Brown, 1981

Odell, P.R., *Oil and World Power*, Harmondsworth: Penguin, 1986

Oppenheim, V.H., 'Arab Tankers Move Downstream', in: *Foreign Policy*, no. 23 (Summer 1976), pp. 117-131

Oppenheim, V.H., 'The Past: We Pushed Them', in: *Foreign Policy*, no. 25 (Winter 1976-1977), pp. 24-57

Partij van de Arbeid, Tweede Kamerfractie van, *Het onderzoek naar de rol van de oliemaatschappijen in de oliecrisis van 1973-1974*, Den Haag: Partij van de Arbeid, 1975

'Primary Energy Sources Indigenous to the Community', in: *European Report*, no. 95, December 1, 1973

Peeters, F., *Gezworen vrienden. Het geheime bondgenootschap tussen Nederland en Israël*, Amsterdam: Veen, 1997

Pearson, S., 'Netherlands Foreign Policy and the 1973-1974 Oil Embargo. The Effects of Transnationalism', in: F.L. Grieves (ed.), *Transnationalism in World Politics and Business*, New York: Pergamon, 1979, pp. 114-140

Pinder, J., *European Community. The Building of a Union*, Oxford: OUP, 1991

Quandt, W.B., *Decade of Decisions. American Policy Towards the Arab-Isreali*

Conflict, 1967-1976, Berkely: University of California, 1977

Rosenthal, U. & G.H. Scholten, *Crisis en continuïteit. Economische Zaken, de oliecrisis en andere turbulenties*, Alphen a/d Rijn: Samson, 1977
Roussel, E., *George Pompidou*, Paris: Fayard, 1984
Rusman, P., 'Dilemma's bij de wapenleveranties aan Israël, 1956-1977', in: P. Everts (ed.), *Dilemma's in de buitenlandse politiek van Nederland*, Leiden: DSWO Pers, 1996, pp. 171-180
Rustow, D.A., 'Who Won The Yom Kippur and Oil Wars?', in: *Foreign Policy*, no. 17 (Winter 1974), pp. 166-175

Sampson, A., *The Seven Sisters. The Great Oil Companies & The World They Shape*, New York: The Viking Press, 1975
Schaper, H.A., 'Nederland en het Midden-Oostenconflict: het regeringsbeleid in de jaren 1967-1973', in: *Internationale Spectator*, vol 29, no. 4 (April 1975), pp. 229-242
Schmidt, H., *Menschen und Mächte*, Berlin: Siedler Verlag, 1987
Schneider, S., *The Oil Price Revolution*, Baltimore & London: John Hopkins University Press, 1983
Schendelen, van M.P.C.M., *Regering en parlement in crisistijd. De machtigingswet als probleem van wetsevaluatie*, Alphen a/d Rijn: Tjeenk Willink, 1979
Serre, F. de la, 'L'Europe des Neuf et le conflit Israélo-Arabe', in: *Revue Francaise de Science Politique*, vol. XXIV, no. 4 (Aout 1974), pp. 801-810
Seumeren, van H., *Gerrit A. Wagner. Een loopbaan bij de Koninklijke*, Utrecht/Antwerpen: Veen, 1989
Sheenan, E.R.F., *The Arabs, Israelis and Kissinger. A Secret History of American Diplomacy in the Middle East*, New York: Reader's Digest Press, 1976
Sheehan, E.R.F., 'How Kissinger did it. Step by Step in the Middle East', in: *Foreign Policy*, no. 22 (Spring 1976), pp. 3-70
Sherrill, R., *The Oil Follies of 1970-1980*, New York: Doubleday, 1983
Simonet, H., 'Energy and the Future of Europe', in: *Foreign Affairs*, vol. 53, no. 1 (April 1975), pp. 450-463
Simonian, H., *The Privileged Partnership. Franco-German Relations in the European Community 1969-1984*, Oxford: Clarendon Press, 1985
SIPRI, *Oil and Security*, Stockholm: Almqvist & Wiksell, 1974
Soetendorp, R.B., *Het Nederlandse beleid ten aanzien van het Arabisch-Israëlisch conflict 1947-1977*, Groningen, 1982
Statistisk årbok Danmark 1972, no. 77, København: Danmark Statistik, 1973
Statistisk årbok Danmark 1973, no. 78, København: Danmark Statistik, 1974
Statistisk årbok Danmark 1974, no. 79, København: Danmark Statistik, 1975

Stemerdink, B., *Tussen dromen, daden en twijfels: dagboekaantekeningen over de sociaal-democratie in Nederland en daarbuiten 1986-1993*, Amsterdam: Van Gennep, 1993

Stuart, D. & W. Tow, *The Limits of Alliance.* NATO *Out-of-Area Problems Since 1949*, Baltimore: John Hopkins University Press, 1990

Sutcliffe, A., *An Economic and Social History of Western Europe Since 1945*, London and New York: Longman, 1996

Swann, D., *The Economics of the Common Market*, London: Penguin, 1990

Szyliowicz J.S. & B.E. O'Neill (ed.), *The Energy Crisis and U.S. Foreign Policy*, New York: Praeger, 1975

Terlingen J. & K. Roskam, *Mahmoud Rabbani. Palestijn in Nederland*, Amsterdam/Antwerpen: Veen, 1997

The Middle East and The International System. Part I: The Impact of 1973 War, London: Aldelphi Paper no. 114, 1975

The Middle East and The International System. Part II: Security and the Energy Crisis, London: Aldelphi Paper no. 115, 1975

Toirkens, J., *Schijn en werkelijkheid van het bezuinigingsbeleid 1975-1986*, Deventer: Kluwer, 1988

Turner, L., *Oil Companies in the International System*, London: Allen & Unwin, 1980

Urwin, D.W., *The Community of Europe. A History of European Integration since 1945*, London & New York: Longman, 1995

Vernon, R. (ed.), *The Oil Crisis*, New York: Norton, 1976

Versloot, H., 'Overgang naar een nieuw energietijdperk', in: *Economisch Statistische Berichten*, 21 november 1973

Voorhoeve, J.J.C., *Peace, Profits and Principles. A Study of Dutch Foreign Policy*, Leiden: Nijhoff, 1985

Wilson, H., *Final Term. The Labour Government 1974-1976*, London: Weidenfeld & Nicolson, 1979

Winkler Prins Jaarboek 1974. Het Jaar in Woord en Beeld, Amsterdam/Brussel: Elsevier, 1974

Yergin, D., *The Prize. The Epic Quest for Oil, Money and Power*, London: Simon and Schuster, 1991

Newspapers

Algemeen Dagblad
Elsevier
De Groene Amsterdammer
Intermediair
International Herald Tribune
Le Monde
NRC *Handelsblad*
Het Parool
De Telegraaf
The Times
Trouw
de Volkskrant
Vrij Nederland

Index of Names

Index of Subjects

www.ingramcontent.com/pod-product-compliance
Ingram Content Group UK Ltd.
Pitfield, Milton Keynes, MK11 3LW, UK
UKHW050051200926
475702UK00019B/2776
9789053564851